Local Density Theory of POLARIZABILITY

PHYSICS OF SOLIDS AND LIQUIDS

AMORPHOUS SOLIDS AND THE LIQUID STATE
Edited by Norman H. March, Robert A. Street, and Mario P. Tosi

CHEMICAL BONDS OUTSIDE METAL SURFACES
Norman H. March

CRYSTALLINE SEMICONDUCTING MATERIALS AND DEVICES
Edited by Paul N. Butcher, Norman H. March, and Mario P. Tosi

ELECTRON SPECTROSCOPY OF CRYSTALS
V. V. Nemoshkalenko and V. G. Aleshin

FRACTALS
Jens Feder

INTERACTION OF ATOMS AND MOLECULES WITH SOLID SURFACES
Edited by V. Bortolani, N. H. March, and M. P. Tosi

LOCAL DENSITY THEORY OF POLARIZABILITY
Gerald D. Mahan and K. R. Subbaswamy

MANY-PARTICLE PHYSICS, Second Edition
Gerald D. Mahan

ORDER AND CHAOS IN NONLINEAR PHYSICAL SYSTEMS
Edited by Stig Lundqvist, Norman H. March, and Mario P. Tosi

THE PHYSICS OF ACTINIDE COMPOUNDS
Paul Erdös and John M. Robinson

POLYMERS, LIQUID CRYSTALS, AND LOW-DIMENSIONAL SOLIDS
Edited by Norman H. March and Mario P. Tosi

THEORY OF THE INHOMOGENEOUS ELECTRON GAS
Edited by Stig Lundqvist and Norman H. March

Local Density Theory of
POLARIZABILITY

Gerald D. Mahan

University of Tennessee
Knoxville, Tennessee

and

K. R. Subbaswamy

University of Kentucky
Lexington, Kentucky

SPRINGER SCIENCE+BUSINESS MEDIA, LLC

Library of Congress Cataloging-in-Publication Data

Mahan, Gerald D.
Local density theory of polarizability / Gerald D. Mahan and K.R. Subbaswamy.
p. cm. -- (Physics of solids and liquids)
Includes bibliographical references and index.
ISBN 978-1-4899-2488-9 ISBN 978-1-4899-2486-5 (eBook)
DOI 10.1007/978-1-4899-2486-5
1. Polarizability (Electricity) 2. Solid state physics. 3. Solid state chemistry. 4. Ionic crystals. I. Subbaswamy, K. R. II. Title. III. Series.
QC173.4.P64M34 1990
530.4'1--dc20 90-42552
CIP

Originally published by Plenum Press, New York in 1990
Softcover reprint of the hardcover 1st edition 1990

Preface

During the past decade the theoretical physics community has learned how to evaluate accurately polarizabilities and susceptibilities for many-electron systems such as atoms, solids, and liquids. The most accurate numerical technique employs a method often called the *Time-Dependent Local Density Approximation*, which is abbreviated TDLDA.

The present volume is a review of recent research on the theory of polarizabilities and susceptibilities. Both authors have been doing these calculations. However, this review surveys the entire field, summarizing the research of many contributors.

The application of an external field, either ac or dc, will induce a dipole moment which can be calculated and compared with experiment. For moderately strong fields, both linear and nonlinear processes contribute to the moment. We cover topics such as polarizability, hyperpolarizability, photoionization, phonons, and piezoelectricity.

Density functional theory in the *Local Density Approximation* (LDA) has been shown to be a very accurate method for calculating ground state properties of electronic system. For static external fields, the induced moments are properties of the ground state. Then the calculation of the polarizability is very accurate. For ac fields, the moment is not part of the ground state. However, the TDLDA methods are still very accurate.

The book is divided into three parts. The first two chapters briefly discuss the LDA formalism: its history, development, and present status. Chapters 3 and 4 discuss the method of calculating linear and nonlinear polarizabilities for atoms. Most of the applications are for atoms and ions with closed shells. The last two chapters discuss the extension to solids, liquids, and surfaces. The latter subjects comprise most of the recent applications and research.

One unique feature of the book is the discussion of numerical techniques. The formulas are first derived algebraically. Then the numerical steps for evaluating the formulas are discussed in detail. Our goal is to make the method accessible to advanced students and researchers. Enough detail is presented to enable them to write their own computer codes and to use the techniques to calculate accurate numbers for real systems. Historically two different, but equivalent numerical methods were developed for the polarizability: the Green's function method and the modified Sternheimer method. We present both techniques. We have done both, and find they are numerically similar, in both the programing time and the CPU time.

We thank Plenum Publishing Corporation for letting us write the manuscript in LaTeX, which permitted us to have control over the appearance and format of the final copy. One of us (KRS) is grateful to Drs. Gaetano Senatore and Michael Johnson for fruitful collaboration on the subject matter of this book. We also thank our families for their patience and understanding during the long winter evenings of writing and rewriting.

Jerry and Swamy

Contents

Chapter 1

Introduction

Electric and magnetic polarizabilities are general response functions for atoms and solids. An experimentalist might impose an external field upon the sample under investigation. The sample could be a collection of atoms in a gas, liquid, or solid. The sample develops an induced moment in response to the field. For example, a spherically symmetric atom would develop an induced dipole moment p in response to an external electric field $\mathcal{E}$ (Buckingham, 1967),

$$p = \alpha\mathcal{E} + \frac{\gamma}{3!}\mathcal{E}^3 + \ldots, \tag{1.1}$$

where α and γ are polarizabilities. The linear polarizability is α and the nonlinear polarizability is γ.

This book discusses methods for calculating polarizabilities such as α and γ. Linear and nonlinear polarizabilities are calculated for atoms, liquids, solids, and surfaces. Many calculations have been done by numerous workers, and these will be reviewed as much as possible.

The systems under investigation each have many electrons. Linear and nonlinear susceptibilities can be calculated exactly for one-electron systems such as the hydrogen atom or the harmonic oscillator. No exact methods are available for many-electron systems. Two different theoretical methods are widely used to calculate polarizabilities for many-electron systems. One is mostly used by chemists, and is called the *Time-Dependent Hartree-Fock Approximation*, which is abbreviated *TDHFA*. This title is a bit misleading, since the equations are not time dependent!

The second method of solving systems with many electrons is called the *Local Density Approximation*, which is abbreviated *LDA*. This method is the focus of this book. We regard it as the easiest numerically, and also the most accurate. The specific extension of the method which is used to calculate the polarizability is called the *Time-Dependent Local Density Approximation* or *TDLDA*. The subject of this book is the calculation of linear and nonlinear polarizabilities using LDA. Polarizabilities are also called susceptibilities.

LDA has a long history. The formal development started with the papers of Hohenberg and Kohn (1964), and Kohn and Sham (1965). Their ideas have been developed into a successful formalism which has been used to calculate a wide variety of phenomena. Calculations are done for atoms, molecules, surfaces, and solids. An excellent review is provided in *Theory of the Inhomogeneous Electron Gas* edited by Lundqvist and March (1983). The calculation of polarizabilities represents a small fraction of the worldwide activity using LDA. Professor L.J. Sham remarked in a lecture in 1989 that he had asked a computer information retrieval system to list all of the papers published using LDA. The response overloaded the system, causing the computer to crash.

The major problem in any calculation for realistic systems is to treat electron-electron interactions. The long range of the Coulomb interaction ensures that each electron interacts with many others in its neighborhood. Condensed-matter physicists and chemists have learned to calculate the properties of many-electron systems in a realistic fashion. This challenge has been met and solved. LDA was developed to provide an accurate and relatively easy technique for calculating the properties of many-electron systems. Before discussing this method, it is useful to first review other methods which preceded it historically.

As a first step, it is useful to state the problem carefully. A system of N electrons interact with each other by Coulomb interactions. They each interact with a fixed potential $U(\vec{r})$. In an atom U is the potential of the nucleus, while for a solid it could be the periodic potential of the ion cores. The Hamiltonian for the N-electron system is

$$H = \sum_j \left(\frac{p_j^2}{2m} + U(\vec{r}_j) \right) + \frac{e^2}{2} \sum_{i,j} \frac{1}{|\vec{r}_i - \vec{r}_j|}. \tag{1.2}$$

This form of the Hamiltonian is suitable for nonrelativitistic interactions. Virtually all of the polarizability calculations which have been done using TDLDA have used nonrelativistic kinematics. This Hamiltonian provides a rigorous starting point for the calculation of many-electron properties. Unless N is a very small number, the Hamiltonian is too complicated to solve directly. For general values of N, one accepts from the outset that approximations have to be made in solving the Hamiltonian.

Virtually all of the approximations to the N-particle Hamiltonian have been aimed at constructing an accurate Hamiltonian for a single electron. In the history of quantum mechanics, Hartree (1928) suggested an approximation which is simple computationally and is also intuitively attractive. His suggestion, which is now called the *Hartree approximation*, was that an effective potential energy for the electron is determined by the average motion of the other electrons:

$$H\psi_\lambda = E_\lambda \psi_\lambda, \tag{1.3}$$

$$H = \frac{p^2}{2m} + U(\vec{r}) + V_H(\vec{r}), \tag{1.4}$$

$$V_H(\vec{r}) = e^2 \int \frac{d^3r' n(\vec{r}')}{|\vec{r} - \vec{r}'|}. \tag{1.5}$$

The quantity $n(\vec{r}')$ is the density of electrons in the system. Choosing this quantity presents problems. The easiest choice, numerically, is to make $n(\vec{r}')$ the total density of all of the electrons. This density includes the contribution of the electron whose wave function ψ_λ is being calculated. Then one is including a particle interacting with itself. This term should not be present, so this choice is an approximation to the true electrostatic potential. In localized systems such as atoms, this choice can make a large error. An electron in a filled p-shell sees the Coulomb potential of five other electrons in the same shell, not six. Including the self-interaction makes an error of 16%. On the other hand, for electrons in the conduction band of a solid, the error from self-interaction is of order $O(1/N)$ which is negligible. The other option is to remove the self-interaction from the electron density $n(\vec{r})$. This removal can be achieved by defining the density n_λ of other electrons,

$$n_\lambda(\vec{r}) = n(\vec{r}) - |\psi_\lambda(\vec{r})|^2 . \tag{1.6}$$

The density n_λ is used inside the Coulomb integral in the definition of the electron-electron potential V_H in (1.5). Using this form for the electron density removes the self-interaction from the Hartree approximation. Unfortunately, it introduces another problem: the potential V_λ for each electron orbital is different. Therefore, the electron orbital states are no longer mutually orthogonal. This nonorthogonality is another nuisance. The Hartree approximation has conceptual problems regardless of how one formulates the potential. Incidently, the LDA also has a similar problem with self-interaction, which is discussed in Chapter 3.

Fock (1930) pointed out that the Hartree approximation neglects an important contribution arising from the anti-symmetry of the many-electron wave function, which is now called *exchange.* He suggested adding another term to the potential. Doing this is now called the *Hartree-Fock Approximation*, which is abbreviated as HFA. In HFA the potential energy for a Hamiltonian of an electron in the eigenstate ψ_λ is

$$V_\lambda(\vec{r})\psi_\lambda(\vec{r}) = (U(\vec{r}) + V_H(\vec{r}))\,\psi_\lambda(\vec{r}) + \sum_{\lambda',\sigma'=\sigma} V_{\lambda\lambda'}(\vec{r})\psi_{\lambda'}(\vec{r}), \quad (1.7)$$

$$V_{\lambda\lambda'} = -e^2 \int \frac{d^3\vec{r}\,'}{|\vec{r}-\vec{r}\,'|}\psi_{\lambda'}(\vec{r}\,')^\dagger\psi_\lambda(\vec{r}\,'). \quad (1.8)$$

The notation $\sigma' = \sigma$ means the sum is only over states with spins which are parallel. The Hartree-Fock potential couples the eigenstate ψ_λ to all of the other eigenstates $\psi_{\lambda'}$. The resulting Schrödinger equation becomes a multidimensional integrodifferential equation, which is also nonlinear. The inclusion of the exchange term increases the accuracy of the potential, while also increasing the computational difficulty.

One calculation which can be done easily is the exchange energy of a homogeneous gas of electrons whose basis states are plane waves. The eigenstates ψ_λ are each assumed to have the form $e^{i\vec{k}\cdot\vec{r}}$, where $\lambda \equiv \vec{k}$. The exchange energy of an electron with wave vector $\vec{k}$ is (see, e.g., Mahan, 1990)

$$\Sigma_x(\vec{k}) = -\frac{1}{\Omega}\sum_{\vec{q}} n_{\vec{k}+\vec{q}}v_q, \quad (1.9)$$

$$v_q = 4\pi e^2/q^2. \quad (1.10)$$

The volume of the system is Ω, and $n_{\vec{k}}$ is the electron occupation number for the orbital $\vec{k}$. The discrete summation over $\vec{q}$ can be changed to a continuous integral. Its evaluation gives

$$\Sigma_x(k) = \frac{e^2k_F}{\pi}S(k/k_F), \quad (1.11)$$

$$S(y) = -\left(1 + \frac{1-y^2}{2y}\ln\left|\frac{1+y}{1-y}\right|\right). \quad (1.12)$$

The self-energy $\Sigma_x(k)$ is the property of a single electron in the state $\vec{k}$, due to the interactions with the other electrons. At $y = 0$ one has that $S(0) = -2$, and $\Sigma_x(0) = -2e^2k_F/\pi$. At the Fermi surface, one has $y = 1, S(1) = -1$, and the self-energy is $\Sigma_x(k_F) = -e^2k_F/\pi$. The self-energy of the electron depends upon its wave vector. Another important quantity is the self-energy of the *system* of electrons. It is obtained by averaging the self-energy of each electron over the set of states occupied by the distribution of electrons

$$E_x = \frac{1}{2\Omega}\sum_{\vec{k}} n_{\vec{k}}\Sigma_x(k), \tag{1.13}$$

$$E_x = -\frac{3e^2k_F}{4\pi}. \tag{1.14}$$

The factor of 1/2 occurs because the exchange energy is between pairs of electrons. The factor of 1/2 ensures that each pair is counted only once. These results for the exchange energy will be useful through the book. They are important ingredients in using LDA.

Although the exchange energy is relatively easy to calculate for plane wave states, it is difficult to calculate for more realistic wave functions which are found in atoms and solids. The Hartree-Fock equations are hard to solve whenever the system has more than a few electrons.

After solving the Hartree-Fock equations, one finds that the resulting eigenfunctions and energies are not very good. They predict atoms which are too big, and binding energies which are too small. The electron-electron interactions cause additional energy terms besides those of Hartree and Fock. Wigner (1938) coined the name of *correlation energy* to describe these additional terms. The additional self-energy is denoted $\Sigma_c(k)$ for a single electron. When this is averaged over the properties of all of the electrons, the correlation contribution to the ground state energy is called E_c. This latter quantity plays a crucial role in LDA. A detailed discussion of the exchange and correlation energies is given in Chapter 3.

The many-electron problem can now be stated in a simple way. The Hamiltonian for N electrons is solved by first deriving and then solving a differential equation for each occupied electron orbital. This differential equation has the appearance of a Schrödinger equation, with an effective potential. The effective potential must include the effects of exchange and correlation.

A historically important step was first taken by Slater (1951, 1953). He was calculating the band structure of solids, where the number of electrons $N \sim 10^{23}$. Solving the Hartree-Fock equations is obviously impossible for that many electrons. In order to include exchange in an approximate way, he decided to construct an exchange potential $V_x(\vec{r})$. His starting point was the exchange energy (1.14) of the homogeneous electron gas. The relationship between the Fermi wave vector k_F and the uniform density n_0 is $k_F = (3\pi^2 n_0)^{1/3}$. Slater suggested using the free-electron result for the exchange energy, while replacing the uniform density n_0 by the actual local density $n(\vec{r})$ of electrons in the solid. He used a potential of the form

$$V_x = -\alpha \frac{e^2(3\pi^2 n(\vec{r}))^{1/3}}{\pi}. \tag{1.15}$$

The parameter α is dimensionless. Slater thought one should use twice the ground state energy, so chose $\alpha = 3/2$. Our standard unit of energy is the Rydberg, which is $Ry = e^2/2a_0$, where a_0 is the Bohr radius. Slater's exchange potential was used in a Schrödinger equation:

$$\left[\frac{-\nabla^2}{2m} + V_x + U(\vec{r}) + V_H(\vec{r})\right]\psi_\lambda = \epsilon_\lambda \psi_\lambda, \tag{1.16}$$

$$V_x(\vec{r}) = -2.95[a_0^3 n(\vec{r})]^{1/3} Ry, \tag{1.17}$$

$$n(\vec{r}) = \sum_\lambda n_\lambda |\psi_\lambda(\vec{r})\,|^2, \tag{1.18}$$

where n_λ is the occupancy (0 or 1) of the orbital ψ_λ. This type of calculation acquired the name of Xα. Slater's choice of potential was entirely intuitive and very *ad hoc*. In fact, the calculations gave poor results. Then it became customary to vary the parameter α in order to seek improvement in the numerical output. This improved the results, but it made the potential appear to be arbitrary. Although *ad hoc*, the potential has a qualitative behavior which is correct. The exchange energy is largest in areas of high density, and lowest in areas of low density. So a potential which is proportional to a power of the density is approximately right.

The Xα method became used widely in the 60s and 70s (see, e.g., Slater, 1975). There was a time when it was the only simple method available to simulate exchange. It constitutes an important historical precedent to LDA.

LDA is a numerical method which is quite similar to Xα in practice. One solves a Schrödinger equation with an effective potential $V_{xc}(\vec{r})$ for exchange and correlation. It has two terms. The first is an exchange term which is just the Slater potential with $\alpha = 1$. The second term is a correlation potential. Each is a local function of the electron density $n(\vec{r})$. The starting equations are only slightly different than those proposed by Slater.

The major difference between Xα and LDA is in the formal underpinning. While Xα is *ad hoc*, LDA is based upon a fundamental theorem. Hohenberg and Kohn (1964) proved that there does exist an exchange-correlation energy which is a *functional of the electron density.* Later, Kohn and Sham (1965) showed how to construct a single-particle exchange-correlation potential from it. One can do exact calculations if one finds first the exact form for V_{xc}. The importance of the theorem is that it ended the discussion over whether the idea of a local potential is valid. Instead, one could focus the attention on finding the best potential.

There is a difference between a *functional* of the density and a *function* of the density. A function is anything such as $n(\vec{r})^{1/3}$ or else $\ln(n(\vec{r}))$. A functional of the density could have $n(\vec{r}')$ as part of the integrand of an integral over d^3r'. The theorem of Hohenberg and Kohn proved there existed a functional of the density. Most numerical solutions employ a function of the density. A lot of effort has gone into finding the most accurate local function.

This book is about polarizabilities and related quantities. The polarizability of an atom can be calculated exactly only for one-electron atoms such as hydrogen. Approximate methods are required in all other cases. Each method of calculating electron states, such as Hartree-Fock, Xα, or LDA, can also be used to calculate the polarizability. Methods to calculate the polarizability were developed for Hartree-Fock and Xα before the era of LDA. A very important idea is that the electric field applied to an atom is screened by the electrons in the outer shells. The screening reduces the field inside of the atom, and also the polarizability. A variety of names are applied to the calculation of this screening: *self-consistent field, time-dependent*, or *coupled equations.* For example, Ahlberg and Goscinski (1973) developed the *time-dependent Xα approximation*, which was a historical precedent to TDLDA. Some of their numerical results are quite good when compared to recent LDA results. These comparisons will be made in Chapter 4.

The polarizability and hyperpolarizability of an atom are important quantities. They arise in a variety of experiments and situations and can be measured. They are also important in condensed matter physics. They influence the static, acoustical, and optical properties of solids and liquids.

The techniques we discuss for the calculation of the polarizability can be applied to many sytems. They have been used, e.g., for atoms, ionic solids, metal spheres and metal surfaces.

Chapter 2

Formalism

2.1 Density Functional Theory

Density function theory is quite widely used. Many reviews and monographs exist on this important topic. Some recent reviews are those by Lundqvist and March (1983), Callaway and March (1984), and Jones and Gunnarsson (1989). They provide more rigor and detail than the present discussion.

Density functional theory is based upon two mathematical theorems. Both were proved in 1964 by Hohenberg and Kohn. The two theorems are:

1. The electron density $n(\vec{r})$ in the ground state is a functional of the potential $U(\vec{r})$.

2. The potential $U(\vec{r})$ is a unique functional (to within a constant) of the density $n(\vec{r})$.

Together these theorems are equivalent to the assertion that *the ground state energy E_G is a minimum with respect to variations in the electron density $n(\vec{r})$.*

Most readers are familiar with variational calculations in quantum mechanics. There the variation is on the wave function, and only applies to the wave function of the ground state. The theorems of Hohenberg and Kohn are similar in spirit but different in the mathematical statement. Note that the electron density is the absolute square of the wave function, and contains no phase information. The density can, in principle at least, be measured experimentally.

The theorems of Hohenberg and Kohn can be made more precise by introducing the concept of *functional derivative.* The ground state energy

E_G is a scalar constant which characterizes the entire system. The density $n(\vec{r})$ depends upon the position $\vec{r}$. The functional derivative of the ground state energy with respect to the electron density is denoted as $\delta E_G/\delta n(\vec{r})$. It is a function of $\vec{r}$, and gives the change in ground state energy at each point as a function of variations in density at that point. This qualitative definition is illustrated by a mathematical definition. First, we give a functional F which has a linear and quadratic dependence upon the density, and then take its functional derivative:

$$F = \int d^3r n(\vec{r})U(\vec{r}) + \int d^3r_1 d^3r_2 n(\vec{r}_1)V(\vec{r}_1 - \vec{r}_2)n(\vec{r}_2), \tag{2.1}$$

$$\frac{\delta F}{\delta n(\vec{r})} = U(\vec{r}) + 2\int d^3r' V(\vec{r} - \vec{r}')n(\vec{r}'). \tag{2.2}$$

Note the factor of 2 in the second term, which occurs because this term in F has two factors of density. The functional derivative is a function of position. This definition is clear whenever the density occurs as a single or double power in the integrand. One can extend the definition to higher powers of the density.

The total number of electrons in a system is usually fixed at some number N. Variations with respect to the density are constrained by the requirement that the total number of electrons is unchanged. Variations with constraints are handled by introducing Lagrange multipliers.

The functional derivative of the ground state energy with respect to density equals a constant which is a Lagrange multiplier. This constant is actually the chemical potential μ:

$$\mu = \frac{\delta E_G}{\delta n(\vec{r})}, \tag{2.3}$$

$$\int d^3r \delta n(\vec{r}) = 0. \tag{2.4}$$

Equation (2.3) is very interesting. The left-hand side is a constant μ. The right-hand side appears to be a function of position $\vec{r}$. Since the two sides are equal, the right-hand side must also be a constant. The electron density of the ground state has the interesting property that taking its functional derivative of the ground state energy gives a constant. Several times we have calculated this functional derivative by numerical methods. It is usually a smooth function but never an absolute constant. That does not disprove the theorem, but only shows that our density $n(\vec{r})$ was not chosen very well.

What sort of terms are included in the ground state energy? In the absence of external fields, we shall consider five different contributions:

1. Kinetic energy
2. Potential energy from the "external" potential $U(\vec{r})$
3. Hartree energy from the potential $V_H(\vec{r})$
4. Exchange energy from the potential $V_x(\vec{r})$
5. Correlation energy from the potential $V_c(\vec{r})$

The second and third items were included in the functional F given in (2.1). Let us try to construct a functional for the kinetic energy. We know in quantum mechanics that the kinetic energy for a system of noninteracting particles is given by the expression

$$T_s = \frac{\hbar^2}{2m} \int d^3r \sum_j \vec{\nabla}\psi_j^\dagger \cdot \vec{\nabla}\psi_j. \tag{2.5}$$

How is this expression represented by the density? The theorem of Hohenberg and Kohn states that there must be a way to represent the ground state energy, including the kinetic energy, in terms of a functional of the density. No one has yet found a good way to represent the kinetic energy alone as a simple functional of the density.

One method is to use the Thomas-Fermi approximation (see, e.g., March, 1975). For a gas of noninteracting fermions, the average kinetic energy in the ground state is

$$KE = \frac{3}{5}E_F = \frac{3\hbar^2 k_F^2}{10m}. \tag{2.6}$$

For unpolarized spin one-half fermions with a uniform density n_0, the Fermi wave vector is given by $k_F = (3\pi^2 n_0)^{1/3}$. The Thomas-Fermi approximation assumes this relation is also valid for nonuniform systems:

$$k_F(\vec{r}) = (3\pi^2 n(\vec{r}))^{1/3}, \tag{2.7}$$

$$KE(\vec{r}) = \frac{3}{5}(3\pi^2 n(\vec{r}))^{2/3} Ry, \tag{2.8}$$

where the unit of energy Ry is the Rydberg. This expression for the kinetic energy is a function of position. Its contribution to the ground state energy is obtained by integrating this expression over the volume, while weighting each point with the particle density:

$$T_s = \int d^3rn(\vec{r})KE(\vec{r}), \tag{2.9}$$

$$= \frac{3}{5}[3\pi^2]^{2/3}\int d^3rn(\vec{r})^{5/3}Ry. \tag{2.10}$$

From now on the symbol Ry will not be written. However, all of our energy units are Rydbergs, and all of the distances are given in terms of the Bohr radius.

Our first attempt at constructing a functional of the ground state energy is to add the kinetic energy term to the quantity F above:

$$E_{TF} = \frac{3}{5}[3\pi^2]^{2/3}\int d^3r[n(\vec{r})]^{5/3} + \int d^3rn(\vec{r})U(\vec{r}) + \int d^3r_1d^3r_2\frac{n(\vec{r}_1)n(\vec{r}_2)}{|\vec{r}_1-\vec{r}_2|}. \tag{2.11}$$

Now take the functional derivative with respect to the density, and set this expression equal to the chemical potential,

$$\mu = [3\pi^2n(\vec{r})]^{2/3} + U(\vec{r}) + 2\int d^3r'\frac{n(\vec{r}')}{|\vec{r}-\vec{r}'|}, \tag{2.12}$$

which can be solved to give the density as a function of the screened potential:

$$n(\vec{r}) = \frac{1}{3\pi^2}[\mu - U(\vec{r}) - V_H(\vec{r})]^{3/2}, \tag{2.13}$$

$$V_H(\vec{r}) = 2\int d^3r'\frac{n(\vec{r}')}{|\vec{r}-\vec{r}'|}. \tag{2.14}$$

The next equation for this derivation is Poisson's equation:

$$\nabla^2V_H(\vec{r}) = -8\pi n(\vec{r}), \tag{2.15}$$

$$= \frac{8}{3\pi}[\mu - U(\vec{r}) - V_H(\vec{r})]^{3/2}. \tag{2.16}$$

This last result is the famous Thomas-Fermi equation. This result is rather old, and numerically not very accurate.

In order to improve the result, the kinetic energy term must be done better. Improvements can be achieved by introducing terms which involve the gradient of the density. This 'gradient expansion' has been investigated extensively (von Weizsäcker, 1935). It is not discussed here since the improvements in the theory have resulted from other terms. The modern theory uses the form for the kinetic energy in terms of wave functions as given in (2.5). Additional improvements in the ground state energy are achieved by introducing terms which account for the effects of exchange and correlation.

2.2 Local Density Approximation

The prior section discussed the density functional theory of Hohenbeg and Kohn. The second landmark paper on this topic was by Kohn and Sham (1965). They introduced several more features of the theory which brought it to the form which is used today. Their theory is called the 'Local Density Approximation,' which is abbreviated LDA.

Their paper made major contributions to two important terms in the equation. One is the kinetic energy term, which they called $T_s[n]$, and the other is the exchange-correlation term, which they called $E_{xc}[n]$. These are added to the potential and Hartree terms which we have been calling F. The equation analogous to (2.11) is

$$E_G = T_s + E_{xc} + F, \tag{2.17}$$

$$F = \int d^3r n(\vec{r})U(\vec{r}) + \int d^3r_1 d^3r_2 \frac{n(\vec{r}_1)n(\vec{r}_2)}{|\vec{r}_1 - \vec{r}_2|}. \tag{2.18}$$

The functional derivative of F is the last two terms in (2.12). Kohn and Sham suggested a technique for treating T_s and E_{xc} which works well and is the basis for LDA.

First we discuss E_{xc} since it is easier. Let $\varepsilon_{xc}(n_0)$ be the *ground state energy per electron from exchange and correlation in an electron gas of uniform density* n_0. In an electron gas of constant density the total ground state energy from exchange-correlation is $E_{xc} = N\varepsilon_{xc}(n_0)$ where N is the total number of electrons. Another way to express N is the integral over all volume of the density n_0. For an electron gas of uniform density the exchange-correlation energy of the ground state is

$$E_{xc} = \int d^3r\, n_0 \varepsilon_{xc}(n_0). \tag{2.19}$$

Kohn and Sham suggested that the inhomogeneous electron gas be treated by replacing, in the above expression, the constant electron density n_0 by the actual electron density $n(\vec{r})$,

$$E_{xc} = \int d^3r n(\vec{r}) \varepsilon_{xc}(n(\vec{r})). \tag{2.20}$$

This assumption is the basis of the LDA version of density functional theory. This approximation should be very good when the density $n(\vec{r})$ is slowly varying in space. In practice, it seems to be very good even in atoms, solids, and at surfaces where the density varies rapidly. The idea of taking the expression $\varepsilon_{xc}(n_0)$ from the homogeneous electron gas, and just replacing the constant density n_0 by the actual density $n(\vec{r})$ is similar to the idea of Slater. However, the Kohn-Sham method is a great improvement because they cast their potentials as due to functional derivatives of the ground state energy terms.

As an example, consider the exchange term alone. According to (1.14) the ground state exchange energy per electron in atomic units is

$$\varepsilon_x = -\frac{3}{2\pi}[3\pi^2 n(\vec{r})]^{1/3}. \tag{2.21}$$

The ground state energy from exchange, and its functional derivative are

$$E_x = -\frac{3}{2\pi}(3\pi^2)^{1/3} \int d^3r n(\vec{r})^{4/3}, \tag{2.22}$$

$$V_x(\vec{r}) \equiv \frac{\delta E_x}{\delta n(\vec{r})} = -\frac{2}{\pi}[3\pi^2 n(\vec{r})]^{1/3}. \tag{2.23}$$

This expression is rather interesting. In the previous section we showed that the exchange energy for an electron at the Fermi surface is $\Sigma_x(k_F) = -e^2 k_F/\pi = -2[3\pi^2 n(\vec{r})]^{1/3}/\pi$ in atomic units. The functional derivative gives a potential V_x which is identical to the electron self-energy $\Sigma_x(k)$ evaluated at the Fermi surface $k = k_F$, where the Fermi wave vector is determined by the local density $k_F = [3\pi^2 n(\vec{r})]^{1/3}$. There is a general theorem due to Seitz (1940) which states that the energy of an electron at the chemical potential is obtained exactly from the functional derivative of the total ground state energy. The exchange energy obeys this theorem. In the Thomas-Fermi approximation (2.13), the energy terms are evaluated for an electron on the Fermi surface. LDA has this same feature.

The above derivation shows that the proper choice of α in the Xα method of Slater is $\alpha = 1$. Slater made the wrong choice for α because he thought the exchange potential of a single electron was derived from its average ground state energy. Instead, it is derived from the self-energy of an electron at the

Fermi surface. Since the Xα method neglected the correlation energy, even the choice of $\alpha = 1$ did not give very good results.

In LDA the exchange and correlation contributions are treated as separate:

$$E_{xc} = E_x + E_c, \quad (2.24)$$
$$V_{xc} = V_x + V_c, \quad (2.25)$$
$$V_c = \frac{\delta E_c}{\delta n(\vec{r})}. \quad (2.26)$$

The exchange potential V_x is the Slater form with $\alpha = 1$, which is given in (2.23). The correlation potential V_c is derived in Section 3.1. It is obtained from the correlation energy of the homogeneous electron gas.

In summary, the potential terms which are important in LDA are

$$V(\vec{r}) = U(\vec{r}) + V_H(\vec{r}) + V_{xc}(\vec{r}), \quad (2.27)$$

where the last two terms are defined in (2.14) and (2.25). The "external" potential $U(\vec{r})$ is different for each problem. For atoms it is the Coulomb potential from the nucleus. In solids or surfaces it is the periodic pseudopotential representing the ion cores of the atoms. It could include externally applied fields.

The Kohn-Sham treatment of the kinetic energy term is simple. They used the eigenfunction description of (2.5) which, in atomic units is

$$T_s = \int d^3r \sum_j \vec{\nabla}\psi_j^\dagger \cdot \vec{\nabla}\psi_j. \quad (2.28)$$

The ground state energy is the summation of this kinetic energy term plus the potential energy terms discussed above. One tries to find an equation by taking a functional derivative of the ground state energy—in this case the variation is with respect to the Hermitian conjugate of the eigenfunction $\psi_j^\dagger$. For the potential energy terms E_V, this is similar to the functional derivative with respect to the density, since the wave function $\psi_j^\dagger$ enters only through the electron density:

$$E_V = F + E_{xc}, \quad (2.29)$$
$$\frac{\delta E_V}{\delta \psi_j^\dagger} = \frac{\delta E_V}{\delta n(\vec{r})}\frac{\delta n(\vec{r})}{\delta \psi_j^\dagger} = \frac{\delta E_V}{\delta n(\vec{r})}\psi_j, \quad (2.30)$$
$$\frac{\delta T_s}{\delta \psi_j^\dagger} = -\nabla^2\psi_j. \quad (2.31)$$

Again the variation of the eigenfunction must be done with the constraint of conserving the number of particles. This step introduces the Lagrange multiplier λ,

$$\begin{aligned} \frac{\delta}{\delta\psi_j^\dagger}\left[E_G - \lambda\int d^3rn(\vec{r})\right] &= 0, & (2.32)\\ \{-\nabla^2 + U(\vec{r}) + V_H(\vec{r}) + V_{xc}(\vec{r})\}\psi_j &= \lambda_j\psi_j, & (2.33)\\ n(\vec{r}) &= \sum_j |\psi_j(\vec{r})|^2. & (2.34)\end{aligned}$$

The density is obtained by summing over the N lowest eigenfunctions, where N is the total number of electrons. Also note that the subscript j has been added to the Lagrange multiplier λ_j.

The above set of equations constitutes a self-consistent way of solving for the ground state energy. One solves the Schrödinger-like equation (2.33) for the eigenfunction ψ_j for each occupied state j. These are referred to as Kohn-Sham equations. These eigenfunctions are used to calculate the electron density. This density is used to calculate the potential terms. This process is iterated until self-consistency is achieved. The equations are nonlinear, since the potentials depend nonlinearly upon the density. However, the equations are solved easily since there is no mixing between states of different j, as found in (1.7) in the rigorous version of HFA.

The end result of this calculation is two important quantities. One is the electron density $n(\vec{r})$, and the other is the ground state energy E_G. The ground state energy is calculated easily after achieving self-consistency in the eigenfunctions. Since the density is the sum of squares of the eigenfunctions, one can rewrite the ground state energy, after an integration by parts in the kinetic energy, as

$$E_G = \sum_j \int d^3r\psi_j^\dagger\{-\nabla^2 + U(\vec{r}) + \frac{1}{2}V_H(\vec{r}) + \varepsilon_{xc}(\vec{r})\}\psi_j. \qquad (2.35)$$

The kinetic energy term can be evaluated by using the Kohn-Sham equations, (2.33),

$$\begin{aligned} E_G &= \sum_j \lambda_j - \int d^3r_1 d^3r_2 \frac{n(\vec{r}_1)n(\vec{r}_2)}{|\vec{r}_1 - \vec{r}_2|} \\ &+ \int d^3rn(\vec{r})[\varepsilon_{xc}(n(\vec{r})) - V_{xc}(n(\vec{r}))]. \qquad (2.36)\end{aligned}$$

The last term may be simplified by using the definition of the exchange-correlation potential V_{xc} as the functional derivative of the term in the ground state energy:

$$V_{xc} = \frac{\delta}{\delta n}\int d^3rn(\vec{r})\varepsilon_{xc}(\vec{r}) \;=\; \varepsilon_{xc} + n(\vec{r})\left(\frac{\partial\varepsilon_{xc}}{\partial n}\right)_{n=n(\vec{r})}, \qquad (2.37)$$

$$\int d^3rn(\vec{r})[\varepsilon_{xc}(\vec{r}) - V_{xc}(\vec{r})] \;=\; -\int d^3rn(\vec{r})^2\left(\frac{\partial\varepsilon_{xc}}{\partial n}\right)_{n=n(\vec{r})}. \qquad (2.38)$$

Since the exchange-correlation energy ε_{xc} always decreases with increasing density, this last term is positive.

We have used the words 'eigenfunction' to denote $\psi_j(\vec{r})$ and 'eigenvalue' to denote λ_j. It is a mistake to think of these functions as representing wave functions and energies of one-electron states. There is no such thing. The only rigorous Hamiltonian for the many-electron system is the one in (1.2). There is no Hamiltonian defined for a single electron. Therefore, there is no such thing as the wave functions and energy eigenvalues for a single electron. The eigenvalues λ_j *do not* represent energy levels of single electrons. The way to think about these quantities is that ψ_j and λ_j are just quantities we compute while solving the LDA equations. We need to find them in order to obtain the answer, but they do not have a simple physical interpretation. We use the words 'Kohn-Sham eigenfunction' and 'Kohn-Sham eigenvalue' to denote these quantities. The only two quantities which have a rigorous interpretation are the ground state energy E_G, and the electron density $n(\vec{r})$. Both of these quantities can be measured experimentally, at least in principle.

So far no mention has been made of the spin of the electron. For atoms with filled shells, the spin of the electron plays only a minor role in the theory. The LDA equations for filled shells have the form shown in (2.33). For atoms with shells which are only partially filled with electrons, there may be more electrons of one spin configuration than with another. For example, there may be more electrons with spin up than with spin down. In that case the exchange-correlation potential will depend upon the spin of the electron. One has to write separate equations for each spin component of the electrons. These coupled equations, plus the need for exchange-correlation potentials which depend upon spin, make calculations for unfilled shells more complicated. These equations are discussed, e.g., in Lundqvist and March (1983). Most of our calculations will be for filled shells, where the spin dependence is unimportant except for rules about how many electrons to put into an orbital.

The Kohn-Sham equations brought LDA to a form which is simple to calculate, and which produces accurate answers. However, one should always remember that the third word in LDA is 'approximation.' One apparent approximation is to use a kinetic energy term (2.28) which: (i) is derived from independent particle states, and (ii) is not a functional of the density. The theorem of Hohenberg and Kohn is that the entire ground state energy is a functional of the electron density. Each term is not a separate functional of the density. When the kinetic energy term is treated differently, then one may be making an approximation. This point has been discussed extensively in the general references cited above. One view is that little error is made by this approximation, since compensating errors are made in obtaining the exchange-correlation potential. These latter quantities also contain contributions from the kinetic energy.

A second approximation is their use of a local function for exchange and correlation potential. The theorem is that $V_{xc}(\vec{r})$ is a function of the density. Making it a local function is an approximation. All of these approximations are aimed at obtaining good results with simple functions for numerical computing. More accurate functions may not be useful if they involve a significant increase in computing time. A third approximation is to use the homogeneous electron gas as the ansatz for constructing the exchange-correlation potentials for the inhomogeneous electron gas. In spite of these approximations, the method has proven to be quite accurate.

Since the Hohenberg-Kohn-Sham (HKS) derivation of the LDA equations, there have been numerous papers which tried to improve the method. Some of these suggestions do help in certain situations, and are useful. However, most of the suggestions increase the computational labor without an obvious improvement in the accuracy. Most calculations employ the LDA form suggested by HKS, or else the spin-dependent version needed for open shell systems. The more recent developments are reviewed, e.g., in Jones and Gunnarsson (1989).

2.3 Linear and Nonlinear Response

Many experiments in condensed matter physics measure the response of a system to a relatively weak external perturbation. Examples are the measurement of an electrical current in response to an applied electric field, magnetization induced by an applied magnetic field, absorption of incident electromagnetic waves, etc. Linear response means that the measured signal is directly proportional to the intensity of the perturbation. Correspondingly, higher order response functions are defined which are proportional to higher powers of the strength of the perturbation. Of course, such a power series description breaks down if the perturbation is strong enough to cause level crossings or transitions in the system.

This book is mostly about the calculation of the response of a material to an external electromagnetic field. The linear dielectric response is discussed in standard textbooks, and leads to the familiar phenomena of refraction, dispersion, and absorption of electromagnetic waves (see, e.g., Ashcroft and Mermin, 1976). The higher order response leads to a host of interesting phenomena that constitute the field of nonlinear optics which came into prominence since the advent of the laser. A detailed discussion of these phenomena is outside the scope of this book, and we refer the reader to the excellent books by Bloembergen (1965) and Shen (1984). Our discussion here will be restricted to introducing the physical quantities of interest, and the context in which they arise.

2.3.1 Response Functions

Within the semiclassical theory of radiation, the propagation of electromagnetic waves in a medium is governed by Maxwell's equations, supplemented by constitutive relations. In particular, the displacement $\vec{D}$ and the electric field $\vec{\mathcal{E}}$ are related by

$$\vec{D} = \vec{\mathcal{E}} + 4\pi\vec{P}, \tag{2.39}$$

where $\vec{P}$ is the polarization or the induced dipole moment density of the medium. In linear response the induced polarization is proportional to the electric field, and one has the familiar relation

$$\vec{D} = \boldsymbol{\epsilon}\cdot\vec{\mathcal{E}}, \tag{2.40}$$

where $\boldsymbol{\epsilon}$ is the dielectric tensor. The induced polarization may contain nonlinear terms in addition to the linear term, i.e.,

$$\vec{P}(\vec{r},t) = \vec{P}^{(1)}(\vec{r},t) + \vec{P}^{(2)}(\vec{r},t) + \vec{P}^{(3)}(\vec{r},t) + \cdots, \tag{2.41}$$

where the superscripts denote the power of the electric field occurring in any particular term. It should be noted that terms involving the gradients of the electric field and cross terms also occur in the expression of (2.41), in general (Buckingham, 1967). Although they are generally neglected in treatments of nonlinearities, we will show in Chapter 5 that they may be important in processes such as Raman scattering in solids.

The response functions are introduced via the explicit constitutive relations (see, e.g., Flytzanis, 1975)

$$\begin{aligned} P_i^{(n)}(\vec{r},t) &= \int_{-\infty}^{t} dt_1 \cdots \int_{-\infty}^{t} dt_n \int d^3r_1 \cdots \int d^3r_n \\ &\quad \times \Phi_{ij\cdots\alpha}^{(n)}(\vec{r},t;\vec{r}_1,t_1;\cdots;\vec{r}_n,t_n) \\ &\quad \times \mathcal{E}_j(\vec{r}_1,t_1)\cdots\mathcal{E}_\alpha(\vec{r}_n,t_n), \end{aligned} \tag{2.42}$$

where the summation convention on repeated subscripts is implied. The response functions $\Phi_{ij\cdots\alpha}^{(n)}$ are subject to conditions resulting from causality, and from the fact that the polarization must be real if the electric field is real. The susceptibilities are the appropriately symmetrized Fourier transforms of the response functions. They are of the form

$$\chi_{ij\ell\ldots}^{(n)}(\vec{k}\omega;\vec{k}_1\omega_1,\ldots,\vec{k}_n\omega_n) = \frac{1}{n!}\sum_{\mathcal{P}} \kappa_{ij\ell\ldots}^{(n)}(\vec{k}\omega;\vec{k}_1\omega_1,\ldots,\vec{k}_n\omega_n), \tag{2.43}$$

where $\mathcal{P}$ refers to the sum over all terms obtained by permutation of $j\vec{k}_1\omega_1$, $\ell\vec{k}_2\omega_2$, ..., and

$$\begin{aligned} \kappa_{ij\ldots}^{(n)}(\vec{k}\omega;\vec{k}_1\omega_1,\ldots,\vec{k}_n\omega_n) &= \int d\tau \int d^3r \int d\tau_1 d^3r_1 \cdots \int d\tau_n d^3r_n \\ &\quad \times \Phi_{ij\ldots}^{(n)}(\vec{r},\tau;\vec{r}_1,\tau_1;\ldots,\vec{r}_n,\tau_n)\exp(i\vec{k}\cdot\vec{r}) \\ &\quad \times \exp\left[i(\vec{k}\cdot\vec{r}+\vec{k}_1\cdot\vec{r}_1+\cdots) - i(\omega\tau+\omega_1\tau_1+\cdots)\right]. \end{aligned} \tag{2.44}$$

For illustration, consider a monochromatic, long wavelength field of frequency ω. The induced polarization to third order in the electric field may be written as

$$\begin{aligned} \vec{P}(t) &= \vec{P}^{(0)} + \chi^{(1)}(\omega)\cdot\mathcal{E}\cos(\omega t) + \chi^{(2)}(0):\mathcal{E}\mathcal{E} + \chi^{(2)}(2\omega):\mathcal{E}\mathcal{E}\cos(2\omega t) \\ &\quad + \chi^{(3)}(\omega)\vdots\mathcal{E}\mathcal{E}\mathcal{E}\cos(\omega t) + \chi^{(3)}(3\omega)\vdots\mathcal{E}\mathcal{E}\mathcal{E}\cos(3\omega t), + \cdots. \end{aligned} \tag{2.45}$$

The first term is due to the permanent dipole moment, if any. The second term is the familiar linear response, and leads to the standard, linear dielectric function. The third term, which represents a static moment proportional to the square of the electric field, describes optical rectification (see, e.g., Shen, 1984). The fourth term describes second harmonic generation. The fifth term leads to the nonlinearity of the refractive index, while the last term is responsible for third harmonic generation. Correspondingly, if one considers several electromagnetic waves of different frequencies incident simultaneously, a variety of phenomena ranging from three wave mixing to coherent Raman scattering are described by the nonlinear susceptibilities that enter into the induced dipole moment.

While phenomenological descriptions of the optical susceptibilies, which are generalizations of the Lorentz model for the linear dielectric constant, are relatively straightforward to develop (Bloembergen, 1965), microscopic calculations of these quantities are quite formidable. Just as in the case of the isolated ground state, the only system where the calculation may be carried out analytically is that of the hydrogen atom, and even there for a static field. As is well known, the linear polarizability may be evaluated exactly. Some low order nonvanishing nonlinear terms have also been evaluated (Waller, 1926; Sewell, 1949). For all other systems approximations are needed.

Microscopic calculations of susceptibilities are necessary for many reasons. First of all, experimental measurements are often the result of many different microscopic contributions, and to gain physical insight one needs to know the different contributions separately. Secondly, theoretical justification of successful empirical rules regarding susceptibilities leads to the recognition of the limits of applicability of such rules. Finally, such calculations put more stringent tests on various calculational schemes for many-body systems than ground state properties do.

2.3.2 Perturbation Theory

The usual starting point for the description of the response of an atom to an external field is standard time-dependent perturbation theory. While the general theory was worked out in the early days of the development of quantum theory, several aspects pertaining to the secular and normalization terms, which are important for nonlinear susceptibilities, have only been clarified more recently. An excellent review of these developments has been provided by Langhoff *et al.* (1972). Our aim in this section is to show how the linear and nonlinear susceptibilities are derived in standard perturbation theory. In subsequent sections modifications appropriate to LDA are considered.

Consider a system governed by the unperturbed, time-independent Hamiltonian $H^{(0)}(\vec{r})$, and a time-dependent perturbation,

$$H^{(1)}(\vec{r},t) = f(t)V^{(1)}(\vec{r},t), \tag{2.46}$$

where $f(t)$ is an appropriate switching function chosen such that

$$H^{(1)}(\vec{r},t \to -\infty) \to 0. \tag{2.47}$$

Assume that at $t = -\infty$ the system is in a particular, normalized, nondegenerate eigenstate ${\phi_j}^{(0)}(\vec{r})$ of the unperturbed Hamiltonian $H^{(0)}(\vec{r})$, with energy ${E_j}^{(0)}$,

$$\Psi(\vec{r},t \to -\infty) = {\phi_j}^{(0)} \exp\left[-i{E_j}^{(0)}t/\hbar\right] \equiv \Psi^{(0)}(\vec{r},t). \tag{2.48}$$

It is assumed that the total Hamiltonian is Hermitian, and that the perturbation is sufficiently smooth and continuous in time such that the normalization of $\Psi^{(0)}$ implies that the wave function is normalized for all times. The system is governed by the time-dependent Schrödinger equation,

$$[H^{(0)}(\vec{r}) + H^{(1)}(\vec{r},t) - i\hbar(\partial/\partial t)]\Psi(\vec{r},t) = 0. \tag{2.49}$$

We now digress to consider the case of a static perturbation turned on adiabatically. In the above formulation, this is accomplished by choosing a suitably smooth switching function, and passing to the limit $t \to \infty$. In this limit, one introduces the ansatz

$$\Psi(\vec{r},t) = \xi_s(\vec{r}) \exp\left(-iE_s t/\hbar\right), \tag{2.50}$$

where $\xi_s(\vec{r})$ is the normalized eigenfunction of the total Hamiltonian obtained from ${\phi_0}^{(0)}(\vec{r})$, and E_s is the corresponding perturbed eigenvalue. The function ξ_s and the perturbed energy E_s are then expanded in a perturbation series,

$$\xi_s(\vec{r}) = \sum_{n=0}^{\infty} \xi^{(n)}(\vec{r}), \tag{2.51}$$

$$E_s = \sum_{n=0}^{\infty} E^{(n)}. \tag{2.52}$$

The superscript (n) here denotes terms of the order of the nth power of the perturbation $H^{(1)}$. Substitution of these expansions into (2.49) leads to the hierarchy of equations

$$\left[H^{(0)}(\vec{r}) - E^{(0)}\right] \xi^{(0)}(\vec{r}) = 0, \tag{2.53}$$

$$\left[H^{(0)}(\vec{r}) - E^{(0)}\right] \xi^{(1)}(\vec{r}) = \left[E^{(1)} - H^{(1)}\right] \xi^{(0)}(\vec{r}), \tag{2.54}$$

$$\left[H^{(0)}(\vec{r}) - E^{(0)}\right] \xi^{(n)}(\vec{r}) = \left[E^{(1)} - H^{(1)}\right] \xi^{(n-1)}(\vec{r}) + \sum_{k=1}^{n} E^{(k)} \xi^{(n-k)}(\vec{r}), \quad n > 1, \tag{2.55}$$

with

$$E^{(1)} = < \xi^{(0)} \mid H^{(1)} \mid \xi^{(0)} >, \tag{2.56}$$

$$E^{(n)} = < \xi^{(0)} \mid H^{(1)} \mid \xi^{(n-1)} > - \sum_{k=1}^{n-1} E^{(k)} < \xi^{(0)} \mid \xi^{(n-k)} >, \quad n > 1, \tag{2.57}$$

and

$$\sum_{k=0}^{n} < \xi^{(k)} \mid \xi^{(n-k)} >= 0, \quad n > 0. \tag{2.58}$$

It is more customary, however, to introduce a wave function $\phi_s(\vec{r})$ with the so-called intermediate normalization,

$$\Psi(\vec{r}, t) = \phi_s(\vec{r}) < \phi_s \mid \phi_s >^{-1/2} \exp\left(-iE_s t/\hbar\right), \tag{2.59}$$

$$\phi_s(\vec{r}) = \sum_{n=0}^{\infty} \phi^{(n)}(\vec{r}), \tag{2.60}$$

$$< \phi^{(0)} \mid \phi^{(0)} > = 1, \tag{2.61}$$

$$< \phi^{(0)} \mid \phi^{(n)} > = 0, \quad n > 0. \tag{2.62}$$

It is readily seen that the two sets of functions are very simply related; e.g.,

$$\xi^{(0)} = \phi^{(0)}, \tag{2.63}$$

$$\xi^{(1)} = \phi^{(1)}, \tag{2.64}$$

$$\xi^{(2)} = \phi^{(2)} - \frac{1}{2} < \phi^{(1)} \mid \phi^{(1)} > \phi^{(0)}, \tag{2.65}$$

$$\xi^{(3)} = \phi^{(3)} - \frac{1}{2} < \phi^{(1)} \mid \phi^{(1)} > \phi^{(1)} - \frac{1}{2}\left(< \phi^{(1)} \mid \phi^{(2)} > + < \phi^{(2)} \mid \phi^{(1)} >\right) \phi^{(0)}, \quad \text{etc.} \tag{2.66}$$

In calculating expectation values up to different orders, it must be remembered that wave function normalization to the appropriate order must be maintained.

We now return to the discussion of a time-dependent perturbation, namely, the solutions of (2.49). Introducing the perturbation functions $\Psi^{(n)}$ via

$$\Psi(\vec{r},t) = \sum_{n=0}^{\infty} \Psi^{(n)}(\vec{r},t), \tag{2.67}$$

we get the hierarchy of equations for the perturbation functions:

$$\left[H^{(0)}(\vec{r}) - i\hbar(\partial/\partial t)\right] \Psi^{(0)}(\vec{r},t) = 0, \tag{2.68}$$

$$\left[H^{(0)}(\vec{r}) - i\hbar(\partial/\partial t)\right] \Psi^{(n)}(\vec{r},t) + H^{(1)}(\vec{r},t)\Psi^{(n-1)}(\vec{r},t) = 0, \quad n > 1. \tag{2.69}$$

The assumed normalization of $\Psi^{(0)}$ and of Ψ requires that the $\Psi^{(n)}$ satisfy

$$\sum_{k=0}^{n} < \Psi^{(k)} \mid \Psi^{(n-k)} >= 0, \quad n > 0. \tag{2.70}$$

The way the secular terms arise may be seen most easily by examining the static limit of the functions $\Psi^{(n)}(\vec{r},t)$. It is readily shown that in the static limit these perturbed functions remain time-dependent (in contrast to the functions $\xi^{(n)}$ and $\phi^{(n)}$ introduced above), and indeed contain time-divergent terms. These terms do not contribute to expectation values as may be seen by comparing (2.50) and (2.67); the secular terms arise from the expansion of an overall phase factor. A careful and consistent treatment of the corresponding secular terms and normalization terms that arise in the case of a time-dependent perturbation has been given by Langhoff *et al.* (1972). Time-dependent generalizations of (2.53) – (2.55) obtain. Specializing to a harmonic perturbation of frequency ω, $H^{(1)}(\vec{r},t) = H^{(1)}(\vec{r})[\exp(i\omega t) + \exp(-i\omega t)]$, one has

$$\phi^{(1)}(\vec{r},t) = \phi^{(1)}_{+1}(\vec{r})e^{i\omega t} + \phi^{(1)}_{-1}(\vec{r})e^{-i\omega t}, \tag{2.71}$$

$$\phi^{(2)}(\vec{r},t) = \phi^{(2)}_{+2}(\vec{r})e^{2i\omega t} + \phi^{(2)}_{-2}(\vec{r})e^{-2i\omega t} + \phi^{(2)}_{0}(\vec{r}) \text{ etc.} \tag{2.72}$$

with

$$[H^{(0)}(\vec{r}) - E^{(0)} \pm \hbar\omega]\phi^{(1)}_{\pm 1} = [E^{(1)} - H^{(1)}(\vec{r})]\phi^{(0)}, \tag{2.73}$$

$$[H^{(0)}(\vec{r}) - E^{(0)} \pm 2\hbar\omega]\phi^{(2)}_{\pm 2} = [E^{(1)} - H^{(1)}(\vec{r})]\phi^{(1)}_{\pm 1} + E^{(2)}_{\pm 2}\phi^{(0)}, \tag{2.74}$$

etc.

The second- and higher-order energy shifts have several terms corresponding to different frequency terms, with

$$E_{\pm 2}^{(2)} = < \phi^{(0)} \mid H^{(1)}(\vec{r}) \mid \phi_{\pm 1}^{(1)} > \text{etc.} \tag{2.75}$$

The standard approach is to expand the various perturbation functions in the complete set of eigenfunctions of the unperturbed Hamiltonian, $H^{(0)}$. The resulting equations for the expansion coefficients and their solution are discussed in all textbooks of quantum mechanics. For a static perturbation, for instance, if we choose the orbital $\xi^{(0)}(\vec{r}) \equiv \phi_j(\vec{r})$ in (2.63), we find the solution $\xi_j^{(1)}(\vec{r})$ of (2.64) to be

$$\xi_j^{(1)}(\vec{r}) = \sum_{\beta \neq j} \frac{< \phi_\beta \mid H^{(1)} \mid \phi_j >}{E_j^{(0)} - E_\beta^{(0)}} \phi_\beta(\vec{r}) \ . \tag{2.76}$$

We give below the expressions for the first- and one of the higher-order susceptibilities that result from this approach for the case of a long-wavelength monochromatic electromagnetic wave incident on a dilute gas of atoms (see Langhoff *et al.* 1972). The contributions from a single electron in an unperturbed orbital $\phi_j(\vec{r})$ are given:

$$\chi^{(1)}(j;\omega) = -ne^2 \sum_{\beta \neq j} \frac{\vec{r}_{j\beta} v_{\beta j}}{(\omega_{\beta j} \pm \omega)}, \tag{2.77}$$

$$\begin{aligned} \chi^{(2)}(j;0) = & -\frac{ne^3}{4} \sum_{\beta \neq j} \left\{ \frac{2\vec{r}_{j\beta} v_{\beta j} v_{jj}}{\omega_{\beta j}(\omega_{\beta j} \pm \omega)} + \frac{\vec{r}_{jj} v_{j\beta} v_{\beta j}}{(\omega_{\beta j} \pm \omega)^2} \right\} \\ & + \frac{ne^3}{4} \sum_{\beta \neq j} \sum_{l \neq j} \left\{ \frac{2\vec{r}_{j\beta} v_{\beta l} v_{lj}}{\omega_{\beta j}(\omega_{lj} \pm \omega)} + \frac{v_{\beta j} \vec{r}_{\beta l} v_{lj}}{(\omega_{\beta j} \pm \omega)(\omega_{lj} \pm \omega)} \right\}, \end{aligned} \tag{2.78}$$

where n is the number density of atoms,

$$\vec{r}_{kl} = < k^{(0)} \mid \vec{r} \mid l^{(0)} >, \tag{2.79}$$

$$v_{kl} = \vec{r}_{kl} \cdot \hat{\eta}, \tag{2.80}$$

with $\hat{\eta}$ being the unit vector in the direction of the electric field. In these expressions, $\pm$ denotes a summation over terms involving both signs, and $\hbar\omega_{ij} = E_j^{(0)} - E_i^{(0)}$. Note that these expressions involve sums over all intermediate states of the unperturbed system, i.e., an infinite set including both bound states and excited states. For many-electron systems, carrying out this procedure explicitly is impractical, if not impossible. Further, the unperturbed states are usually obtained by some approximation procedure such as coupled Hartree-Fock or local density theory, and are not reliable for the unoccupied levels. The sum over intermediate states can be circumvented either by directly solving for the perturbation functions (2.67) or using a Green's function method. These methods will be discussed in the next chapter.

2.3.3 Static Perturbations

Our goal is to calculate the response of many-electron systems to external perturbations. The many-electron systems can be atoms, ions, solids, surfaces, or liquids. The perturbations can be due to electric fields or other forms of potentials. The formalism for static perturbations is simpler than for ac fields, and is treated separately here for later reference. The external perturbation has a field $\mathcal{E}$. This symbol could denote the electric field, but it also could mean any other perturbation. This field is treated as a scalar quantity: it would be the magnitude of the electric field or any other vector. Each eigenfunction ψ_i is perturbed by the external field. For nonlinear response, consider that the eigenfunctions can be expanded in a power series in this field. The density can also be expanded, since it is determined by the eigenfunctions:

$$\psi_i(\vec{r},\mathcal{E}) = \psi_i(\vec{r}) + \sum_{k=1}^{\infty} \psi_i^{(k)}(\vec{r})\mathcal{E}^k, \tag{2.81}$$

$$n(\vec{r},\mathcal{E}) = n^{(0)}(\vec{r}) + \sum_{k=1}^{\infty} n^{(k)}(\vec{r})\mathcal{E}^k, \tag{2.82}$$

$$n^{(1)}(\vec{r}) = 2\Re\sum_i\{\psi_i^*\psi_i^{(1)}\}, \tag{2.83}$$

$$n^{(2)}(\vec{r}) = \sum_i[|\psi_i^{(1)}|^2 + 2\Re(\psi_i^*\psi_i^{(2)})], \tag{2.84}$$

$$n^{(3)}(\vec{r}) = 2\sum_i^{\infty}\Re[\psi_i^*\psi_i^{(3)} + \psi_i^{(1)*}\psi_i^{(2)}], \tag{2.85}$$

$$n^{(4)}(\vec{r}) = \sum_i\{|\psi_i^{(2)}|^2 + 2\Re[\psi_i^*\psi_i^{(4)} + \psi_i^{(1)*}\psi_i^{(3)}]\}. \tag{2.86}$$

The superscript (k) denotes terms which are kth order in the field. These formulas provide the changes in density which are of various powers of the external field. The external potential V_{ext} is usually linear in the applied field. However, we will be interested in nonlinear potentials, and we assume the intensity of the kth order component is $V_{ext}^{(k)}$.

The next quantity we wish to calculate is the nth-order change in the ground state energy $E_G^{(k)}$,

$$E_G(\mathcal{E}) = E_G^{(0)} + \sum_{k=1} E_G^{(k)} \mathcal{E}^k. \tag{2.87}$$

The density $n(\vec{r}, \mathcal{E})$ and potential $V_{ext}(\vec{r}, \mathcal{E})$ are both expanded in powers of the field. How are these terms combined into an expression for the ground state energy? A simple derivation was provided by DeCicco and Johnson (1969). They started with the theorem of Feynman (1939) which relates the macroscopic force to the average of the microscopic force:

$$\frac{\partial E_G}{\partial \mathcal{E}} = \int d^3r n(\vec{r}, \mathcal{E}) \frac{\partial V_{ext}(\vec{r}, \mathcal{E})}{\partial \mathcal{E}}, \tag{2.88}$$

$$\frac{\partial V_{ext}}{\partial \mathcal{E}} = \sum_{k=1} k V_{ext}^{(k)} \mathcal{E}^{k-1}. \tag{2.89}$$

In the above equation, we have given the expansion for the derivative of the external potential. A similar expansion must exist for the ground state energy,

$$\frac{\partial E_G}{\partial \mathcal{E}} = \sum_{k=1} k E_G^{(k)} \mathcal{E}^{k-1}. \tag{2.90}$$

Comparing this series with the one obtained by expanding (2.88) allows us to identify the terms in the expansion for the ground state energy:

$$E_G^{(1)} = \int d^3r n^{(0)}(\vec{r}) V_{ext}^{(1)}(\vec{r}), \tag{2.91}$$

$$E_G^{(2)} = \int d^3r [n^{(0)}(\vec{r}) V_{ext}^{(2)}(\vec{r}) + \frac{1}{2} n^{(1)}(\vec{r}) V_{ext}^{(1)}(\vec{r})], \tag{2.92}$$

$$E_G^{(3)} = \int d^3r [n^{(0)}(\vec{r}) V_{ext}^{(3)}(\vec{r}) + \frac{2}{3} n^{(1)}(\vec{r}) V_{ext}^{(2)}(\vec{r}) + \frac{1}{3} n^{(2)}(\vec{r}) V_{ext}^{(1)}(\vec{r})], \tag{2.93}$$

$$E_G^{(4)} = \int d^3r [n^{(0)}(\vec{r}) V_{ext}^{(4)}(\vec{r}) + \frac{3}{4} n^{(1)}(\vec{r}) V_{ext}^{(3)}(\vec{r}) + \frac{1}{2} n^{(2)}(\vec{r}) V_{ext}^{(2)}(\vec{r}) + \frac{1}{4} n^{(3)}(\vec{r}) V_{ext}^{(1)}(\vec{r})]. \tag{2.94}$$

These terms can be used to calculate the linear $E_G^{(1)}$ and nonlinear response of the system to the external field. Often we calculate the induced moment $p_{\mathcal{E}}$ due to the field. It is obtained by taking the derivative of the ground state energy,

$$p_{\mathcal{E}} = \frac{\partial E_G}{\partial \mathcal{E}}. \tag{2.95}$$

This expression is just the series in (2.90).

The usual case is to have the external potential depend only on the first power of the field. Then the only term in the external potential is $V_{ext}^{(1)}$ and the terms in the ground state energy have just one term,

$$E_G^{(k)} = \frac{1}{k}\int d^3r n^{(k-1)}(\vec{r})V_{ext}^{(1)}(\vec{r}). \tag{2.96}$$

A specific example is an atom with spherical symmetry in an external electric field, which is in the $\hat{z}$ direction. Then the change in the ground state energy is determined by the dipole polarizability α_d plus higher nonlinear coefficients:

$$V_{ext}^{(1)}(\vec{r}) = -ez\mathcal{E}, \tag{2.97}$$

$$E_G(\mathcal{E}) = E_G^{(0)} - \frac{1}{2!}\alpha_d\mathcal{E}^2 - \frac{1}{4!}\gamma\mathcal{E}^4 - \ldots, \tag{2.98}$$

$$\alpha_d = -e^2\int d^3r\, zn^{(1)}(\vec{r}), \tag{2.99}$$

$$\gamma = -3!e^4\int d^3r\, zn^{(3)}(\vec{r}). \tag{2.100}$$

These formulas will be the basis for the calculation of the polarizability and hyperpolarizability for static external fields.

The external potential is not the only perturbing potential. The electrons will try to screen the external potential. The final potential is called the screened potential. Another name is the *self-consistent field*, and the potential is called V_{SCF}. The screened potential depends upon the density of electrons $n(\vec{r}, \mathcal{E})$. Since this density may be expanded in powers of this field, the SCF potential has a corresponding expansion:

$$V_{SCF}(\vec{r}, \mathcal{E}) = \sum_k V_{SCF}^{(k)}(\vec{r})\mathcal{E}^k. \tag{2.101}$$

Thus, the potential V_{SCF} contains all powers of the external field, although the actual external potential may be only linear in this field. Note that one does *not* use the self-consistent potential in evaluating the integrals (2.96) for the ground state energy, although V_{SCF} is used for the calculation of the density coefficients $n^{(k)}(\vec{r})$. The self-consistent potential enters into the calculation of $\psi_i^{(k)}$. The definition of V_{SCF}, and its role in the calculation of the polarizability, is discussed in the following section.

2.3.4 Self-Consistent Screening

When an external perturbation is imposed on a many-electron system, the electrons screen the field to some extent. In the treatment of Section 2.3.2, this is not taken into account, as evidenced by the occurrence of the "bare" perturbation $H^{(1)}$ on the right hand side of (2.53)–(2.55). In single-particle theories screening may be taken into account by many-body perturbation techniques (see, e.g., Mahan, 1990), and leads to such approximations as the random phase approximation (RPA), etc. In local density theory screening enters via the density dependence of the ground state potential through the terms $V_H(\vec{r})$ and $V_{xc}(\vec{r})$. We restrict our discussion here to the case of a static perturbation. The generalization to a harmonic perturbation follows from the discussion in Section 2.3.2. In the presence of the perturbation $V_{ext}(\vec{r})$, the density changes,

$$n(\vec{r}) \rightarrow n^{(0)} + n_{ind}(\vec{r}), \tag{2.102}$$

and, therefore, both V_H and V_{xc} change:

$$V_H(\vec{r}) + V_{xc}(\vec{r}) \rightarrow V_H^{(0)}(\vec{r}) + V_{xc}^{(0)} + V_{ind}(\vec{r}). \tag{2.103}$$

Thus, (2.49) is modified to

$$[H^{(0)}(\vec{r}) + V_{ext}(\vec{r}) + V_{ind}(\vec{r}) - \lambda]\psi(\vec{r}) = 0. \tag{2.104}$$

The following discussion of screening in LDA is a perturbative treatment of the Kohn-Sham equations, (2.33), regarded as a Hamiltonian problem with an effective potential. Such a treatment would appear to be rather *ad hoc*, and a significant departure from the density functional formalism. A more rigorous justification for this procedure is given in the next section. Here, we merely apply the perturbation theory of Section 2.3.2 to the Kohn-Sham orbitals. Developing the perturbation series, we expand V_{ind} and n_{ind} in powers of the external perturbation. From the definition $n(\vec{r}) = \mid \Psi(\vec{r}) \mid^2$, and (2.49) it is seen that

$$n^{(1)}(\vec{r}) = \sum_{j(occ.)} \left(\psi_j^{(0)^*}(\vec{r})\psi_j^{(1)}(\vec{r}) + \text{c.c.}\right), \quad (2.105)$$

$$n^{(2)}(\vec{r}) = \sum_{j(occ.)} \mid \psi_j^{(1)} \mid^2 + \sum_{j(occ.)} \left(\psi_j^{(0)^*}(\vec{r})\psi_j^{(2)}(\vec{r}) + \text{c.c.}\right), \quad (2.106)$$

etc.

It is convenient to include the external perturbation in the first-order term and write

$$V_{ext} + V_{ind}(\vec{r}) = V^{(1)}(\vec{r}) + V^{(2)}(\vec{r}) + \cdots . \quad (2.107)$$

The hierarchy (2.54) – (2.55) is modified to a hierarchy of Kohn-Sham equations,

$$\left[H^{(0)}(\vec{r}) - \lambda^{(0)}\right] \psi^{(n)}(\vec{r}) = \sum_{k=1}^{n} \left[\lambda^{(k)} - V^{(k)}\right] \psi^{(n-k)}(\vec{r}). \quad (2.108)$$

Correspondingly, the expressions for $\epsilon^{(n)}$, (2.56)–(2.57), are also modified:

$$\lambda^{(1)} = < \psi^{(0)} \mid V^{(1)} \mid \psi^{(0)} >, \quad (2.109)$$

$$\lambda^{(n)} = < \psi^{(0)} \mid V^{(1)} \mid \psi^{(n-1)} > - \sum_{k=1}^{n-1} < \psi^{(0)} \mid \lambda^{(k)} - V^{k)} \mid \psi^{(n-k)} >, \quad n > 1. \quad (2.110)$$

Note that now the perturbed orbitals are coupled because V_{ind} itself depends on all the occupied, perturbed orbitals. Thus, (2.108) requires self-consistent solutions. For this reason, the calculation of susceptibilities while retaining V_{ind} is referred to as *Coupled Hartree-Fock* or CHF approximation. Sometimes, this is also referred to as *Time-Dependent Hartree-Fock* or TDHF approximation. Correspondingly, the evaluation of susceptibilities with V_{ind} self-consistently in LDA is referred to as TDLDA.

From (2.103) it is clear that $V_{ind} = \delta V_H + \delta V_{xc}$. The exact expression for δV_H is given by

$$\delta V_H(\vec{r}) = 2 \int \frac{n_{ind}(\vec{r}\,')}{\mid \vec{r} - \vec{r}\,' \mid} d^3r\,'. \quad (2.111)$$

If we assume that the perturbation is weak, and that the induced change in the density is small, we may attempt a Taylor series for δV_{xc} (Singhal and

Callaway, 1976), i.e.,

$$\delta V_{xc} = \left(\frac{\partial V_{xc}}{\partial n}\right) n_{ind} + \frac{1}{2!}\left(\frac{\partial^2 V_{xc}}{\partial n^2}\right) n_{ind}^2 + \cdots. \tag{2.112}$$

In the above equation, the derivatives are evaluated at the local value of the unperturbed density, which is consistent with the local density approximation made for the unperturbed calculation. Thus, finally, we get

$$V_{SCF}^{(\alpha)}(\vec{r}) = V_{ext}\delta_{\alpha,1} + 2\int \frac{n^{(\alpha)}(\vec{r}\,')}{|\vec{r} - \vec{r}\,'|} d^3r\,' + \delta V_{xc}^{(\alpha)}, \tag{2.113}$$

with

$$\delta V_{xc}^{(1)}(\vec{r}) = \left(\frac{\partial V_{xc}}{\partial n}\right)_{n=n_0(\vec{r})} n^{(1)}(\vec{r}), \tag{2.114}$$

$$\delta V_{xc}^{(2)}(\vec{r}) = \left(\frac{\partial V_{xc}}{\partial n}\right)_{n=n_0(\vec{r})} n^{(2)}(\vec{r}) + \frac{1}{2}\left(\frac{\partial^2 V_{xc}}{\partial n^2}\right)_{n=n_0(\vec{r})} [n^{(1)}(\vec{r})]^2,$$

$$\text{etc.} \tag{2.115}$$

The induced mth order response is obtained via

$$p^{(m)} = \int d^3r V_{ext} n^{(m)}(\vec{r}). \tag{2.116}$$

The equations developed here represent a completely self-consistent scheme to calculate screened linear and nonlinear response to time-dependent perturbations. In passing, we note that if only the induced change in the Hartree potential is included in the screening term, the conventional RPA response is obtained.

2.4 Time-Dependent Potentials in LDA

The perturbation formulation of Section 2.3.2 assumes that the exact eigenvalue and eigenfunction for the unperturbed state are known. Obviously, this is almost never the case. The unperturbed problem is solved in some approximate way, usually using some variational scheme such as Hartree-Fock, Slater Xα, or LDA. In such situations, it is important to obtain a perturbation scheme which is consistent with the variational principle used for the unperturbed problem. For the Hartree-Fock and Slater methods the starting point is the Dirac-Frenkel time-dependent variational principle (Dirac, 1930; Frenkel, 1934),

$$< \delta\tilde{\Psi} \mid H^{(0)} + H^{(1)} - i\hbar(\partial/\partial t) \mid \tilde{\Psi} >= 0, \qquad (2.117)$$

where $\tilde{\Psi}$ is a trial wave function, and $\delta\tilde{\Psi}$ is an arbitrary variation about it. By adding (2.117) to its complex conjugate, we get

$$\delta < \tilde{\Psi} \mid H^{(0)} + H^{(1)} - i\hbar(\partial/\partial t) \mid \tilde{\Psi} > +i\hbar(\partial/\partial t) < \delta\tilde{\Psi} \mid \tilde{\Psi} >= 0. \qquad (2.118)$$

This variational principle, unlike the Ritz principle for the time-independent case, is not in the form of the variation of a functional set equal to zero, and does not provide an upper bound to a property of interest. However, for a harmonic perturbation, a time-averaged functional which provides an upper bound to the energy shift can be obtained as shown by Langhoff *et al.* (1972). If a single Slater determinant is taken as the trial wave function, one is led to the *time-dependent Hartree-Fock* method. The various order perturbed one-electron orbitals obey inhomogeneous Hartree-Fock equations corresponding to the hierarchy (2.53)–(2.55). However, because of the nonlocal nature of the Fock potential, the inhomogeneous terms contain not only the bare external perturbation, but also terms arising from the induced changes in the orbitals to the same order, i.e, terms arising from *exchange screening.* This results in what is known as coupled Hartree-Fock (CHF) equations, which are widely used in calculations involving atoms and molecules (see, e.g., Fowler and Madden, 1985, and references therein).

Ahlberg and Goscinski (1975) considered linear response to time-dependent perturbations in local potential theories of the Xα type. They derived the response from the Dirac-Frenkel variation principle described above, and showed how polarizabilities and shielding factors could be calculated. This work is the precursor for the LDA formulations that are currently popular.

As outlined in Section 2.1, the Hohenberg-Kohn theorem applies rigorously for the case of a static external potential. In the ground state problem, one treats the potential due to the positive ions or nuclei as the external potential. However, it is perfectly valid to include an externally applied static electric field or other static perturbations in the problem. Then, barring any further approximations, one obtains the ground state energy of the system in the presence of the field to the same degree of accuracy as that of the system in the absence of the static external field. This is the basis of the so-called "frozen field" method, which has been successfully applied to obtain not only the dielectric susceptibilities, but also phonon dispersions, Raman tensor, etc. (see, e.g., Devreese *et al.*, 1983).

The Hohenberg-Kohn theorem does not apply for the case of a time-dependent potential. Nevertheless, perturbative methods based on the local density approximation, such as those described in detail in the preceding section, are very successful. Prompted by this, many authors have tried to rationalize these perturbative methods from a more rigorous point of view. In this section we summarize these attempts and the problems they encounter. We note, however, that practical local density schemes for time-dependent perturbations were introduced on an *ad hoc* basis (Zangwill and Soven, 1980; Stott and Zaremba, 1980; Mahan, 1980) long before the more rigorous rationalizations were proposed.

For a system of interacting electrons moving in a static potential $U(\vec{r})$, the Hohenberg-Kohn theorem states that there exists a one-to-one correspondence between the external potential and the ground state electron density $n(\vec{r})$ when the ground state is nondegenerate. There have been extensions of this result for two special cases: (i) periodic potentials (Bartolatti, 1981), and (ii) potentials consisting of a large static part and a small time-dependent part which can be handled in linear response (Chakravarty *et al.*, 1979). The most general result proven to date is the following theorem due to Runge and Gross (1984): *For every single-particle potential that can be expanded into a Taylor series with respect to the time coordinate around* $t = t_0$*, where* t_0 *is the time at which the time-dependent external potential is turned on, the ground state density determines the potential up to a purely time-dependent additive piece.* They also showed that generalization of the Kohn-Sham formulation, where the single-particle orbitals obey time-dependent Schrödinger equations can be obtained. There is some controversy over the mathematical rigor of the proof (Xu and Rajagopal, 1985), but such is also the case with the static formulation. There are also some extensions of the proof (see, e.g., Kohl and Driezler, 1986; Ng and Singwi, 1987).

A local density approximation for the frequency dependent exchange-correlation potential within linear response theory was provided by Gross and Kohn (1985), whose work we summarize. Consider an unperturbed inhomogeneous electronic system with density $n^{(0)}(\vec{r})$ in a nondegenerate ground state of the static external potential $U(\vec{r})$. Let there be a small perturbing potential $V_{ext}(\vec{r},t)$, and a corresponding density response $n^{(1)}(\vec{r},t)$. For linear response, the Fourier components of the two quantities are related by

$$n^{(1)}(\vec{r},\omega) = \int d^3r'\chi(\vec{r},\vec{r}\,';\omega)V_{ext}(\vec{r}\,',\omega), \tag{2.119}$$

where $\chi(\vec{r},\vec{r}\,';\omega)$ is the exact density-density response function. To derive a local density approximation, one assumes that the density $n^{(0)}(\vec{r})+n^{(1)}(\vec{r},t)$ can be reproduced by a system of noninteracting particles in an appropriate single-particle potential $V_{eff}^{(0)}(\vec{r}) + V_{eff}^{(1)}(\vec{r},t)$. Then, one can write

$$n^{(1)}(\vec{r},\omega) = \int d^3r' \chi_{KS}(\vec{r},\vec{r}';\omega) V_{eff}^{(1)}(\vec{r}',\omega), \tag{2.120}$$

where $\chi_{KS}(\vec{r},\vec{r}';\omega)$ is the density-density response function of the noninteracting Kohn-Sham ground state corresponding to $V_{eff}^{(1)}(\vec{r})$, given in terms of the Kohn-Sham orbitals and eigenvalues by

$$\chi_{KS}(\vec{r},\vec{r}';\omega) = \sum_{i,j}(f_i - f_j)\frac{\psi_i(\vec{r})^*\psi_j(\vec{r})\psi_j(\vec{r}')^*\psi_i(\vec{r}')}{\hbar\omega - (\epsilon_j - \epsilon_i) + i\delta}. \tag{2.121}$$

Here, $f_i (= 0 \text{ or } 1)$ are the occupation numbers. By analogy with the static case, one defines the exchange-correlation part of $V_{eff}^{(1)}$ by

$$V_{eff}^{(1)}(\vec{r},\omega) = V_{ext}(\vec{r},\omega) + 2\int \frac{n^{(1)}(\vec{r}',\omega)}{|\vec{r}-\vec{r}'|} d^3r' + \delta V_{xc}^{(1)}(\vec{r},\omega), \tag{2.122}$$

$$\delta V_{xc}^{(1)}(\vec{r},\omega) = \int d^3r' f_{xc}(\vec{r},\vec{r}';\omega) n^{(1)}(\vec{r}',\omega), \tag{2.123}$$

where f_{xc} depends on the unperturbed ground state density. From the above equations, one gets the following formal expression for f_{xc}:

$$f_{xc}(\vec{r},\vec{r}';\omega) = \chi_{KS}^{-1}(\vec{r},\vec{r}';\omega) - \chi^{-1}(\vec{r},\vec{r}';\omega) - 1/|\vec{r}-\vec{r}'|, \tag{2.124}$$

assuming that the inverse response functions exist. For a homogeneous electron gas, the quantity f_{xc} is explicitly known, and the Lindhard expression takes the place of χ_{KS}. It should be noted that for the homogeneous electron gas,

$$f_{xc}^h(\vec{q},\omega) = -(4\pi/q^2)G(\vec{q},\omega), \tag{2.125}$$

where $G(\vec{q},\omega)$ is the local field factor (see, e.g., Mahan, 1990). A local density approximation to the time-dependent problem for the inhomogeneous electron gas is obtained by assuming that (i) $n^{(0)}(\vec{r})$ is sufficiently slowly varying that f_{xc} can be replaced by f_{xc}^h evaluated for the local density $n^{(0)}(\vec{r})$, and (ii) $n^{(1)}(\vec{r}',\omega)$ is sufficiently slowly varying that it may be replaced by $n^{(1)}(\vec{r},\omega)$ in (2.123). This results in

$$\delta V_{xc}^{(1)}(\vec{r},\omega) \approx f_{xc}^h(q=0,\omega;n^{(0)}(\vec{r}))n^{(1)}(\vec{r},\omega). \tag{2.126}$$

The above results provide a basis for the consideration of time-dependent perturbations within LDA. We note here that in the limit $\omega \to 0$ one can show (Kugler, 1975) from the well-known properties of the homogeneous electron gas that

$$\lim_{q \to 0} f_{xc}^{h}(q, 0; n) = \frac{d^2}{dn^2}[n\varepsilon_{xc}(n)], \tag{2.127}$$

where $\varepsilon_{xc}(n)$ has been defined in (2.19). By virtue of (2.126)–(2.127), the quantity $\delta V_{xc}^{(1)}$ defined in (2.123) is identical to the expression derived in (2.109) by *ad hoc* perturbation theory! It is easy to show now that the induced density $n^{(1)}$ derived in Section 2.3.4 tacitly assumes the $\omega \to 0$ limit of f_{xc}^{h}, i.e., that the local response of the electron gas to the time-dependent perturbation is adiabatic. Clearly, this assumption breaks down when the frequency of the applied perturbation is near an electronic resonance of the system. Nevertheless, the scheme gives very accurate results for linear and nonlinear response in a great many situations, as discussed in later chapters.

Chapter 3

Computational Techniques

3.1 Exchange-Correlation Potentials

A key feature of LDA is the use of a potential which is a local function of the density to approximate the effects of exchange and correlation. This local function must be chosen well in order for the theory to be accurate and useful. In this section we summarize some of the many approaches which have been applied to this problem. First we recall the definitions of some of the terms which are used:

- E_{xc} is the total exchange-correlation energy of the system of N electrons.
- $\varepsilon_{xc}(n_0)$ is the average energy per electron from exchange-correlation in a uniform electron gas of density n_0.
- $V_{xc}(\vec{r})$ is the potential energy for a single electron from exchange-correlation.

In LDA the exchange-correlation potential is a functional derivative of the ground state energy contribution from exchange and correlation. The latter is approximated by using the expression for the homogeneous electron gas and replacing the uniform density n_0 by the actual electron density $n(\vec{r})$ (see (2.19)):

$$E_{xc} = \int d^3r n(\vec{r}) \varepsilon_{xc}(n(\vec{r})), \tag{3.1}$$

$$V_{xc}(\vec{r}) = \frac{\delta E_{xc}}{\delta n(\vec{r})} = \left(\frac{\delta(n\varepsilon_{xc}(n))}{\delta n} \right)_{n=n(\vec{r})} \tag{3.2}$$

It is well to remember that this procedure is approximate. The potential V_{xc} is rigorously given by the functional derivative of the ground state energy. However, the ground state energy density is not necessarily a local function of density. What is the exact formula for $\varepsilon_{xc}(n_0)$? Having an accurate expression for this function is essential for the success of the method. The process of numerical computing is made easier by also having this function be simple: either an analytical function or else a simple numerical table. There have been numerous attempts by theorists to calculate the exchange-correlation energy of the homogeneous electron gas. Most calculations obtain ε_{xc} as a function of the uniform density n_0. The scale of density is usually represented by a parameter called r_s. It is the radius in atomic units of a sphere which encloses one unit of electron charge,

$$\frac{4\pi}{3} n_0 a_0^3 r_s^3 = 1, \tag{3.3}$$

where a_0 is the Bohr radius. The parameter r_s is large for an electron gas of low density, while r_s is small for a gas of high density. In metals the range of electron density spans values of r_s from about 1 to 6. This represents a relative variation of of electron density from 1^{-3} to $6^{-3} \approx 0.005$. For atoms and nonmetallic solids the relevant range is much broader. The Fermi wave vector k_F and the exchange energy can be written using r_s:

$$k_F a_0 = (3\pi^2 n_0)^{1/3} a_0 = \left(\frac{9\pi}{4}\right)^{1/3} \frac{1}{r_s}, \tag{3.4}$$

$$\varepsilon_x = -\frac{3}{2\pi} k_F a_0 = -\frac{3}{2\pi}\left(\frac{9\pi}{4}\right)^{1/3} \frac{1}{r_s} = -\frac{0.91632}{r_s}, \tag{3.5}$$

$$V_x = -\frac{2}{\pi}\left(\frac{9\pi}{4}\right)^{1/3} \frac{1}{r_s} = -\frac{1.2218}{r_s}. \tag{3.6}$$

These quantities were defined earlier in (2.21) and (2.23). Most theories report the correlation energy ε_c as a function of the density parameter r_s.

Here we summarize four different theories of ε_c. Each theory is quite ingenious and superficially different. However, the four theories give similar results. They are actually closely related. They represent four different ways of viewing the same problem. The agreement between the various methods means that the problem is solved. The value of ε_c as a function of the density r_s is known to within a few percent. Since ε_c is itself only about 10% of the total ground state energy, then the total ground state energy is known as a function of density to an accuracy of 1%. However, the errors caused by this uncertainty are increased when we take a (functional) derivative of the result, which is required to obtain the effective potential energy for a single electron. The four theories are presented in their historical order. The first two were coincident.

First we present the Singwi-Sjölander result. Vashishta and Singwi (1972) used the method of Singwi and Sjölander (1967) to obtain the following expression for the correlation energy per electron for the unpolarized, homogeneous, electron gas:

$$\varepsilon_c = -0.112 + 0.0335 \ln r_s - \frac{0.027}{0.1 + r_s}. \tag{3.7}$$

The energy is in Rydberg units. Their formula is not obtained by an analytical derivation. Instead, they did a long numerical calculation, obtained the energy at a series of density points, and found empirically that the above expression fitted their points to an accuracy of 1/2%. The formula is valid in the range of density $1 < r_s < 6$. The analytical formula is very useful in LDA for taking the functional derivative in (2.37) which is needed for the correlation potential,

$$V_c = \varepsilon_c - \frac{r_s}{3}\frac{\partial \varepsilon_c}{\partial r_s}, \tag{3.8}$$

$$V_c = -0.123 + 0.0335 \ln r_s - \frac{0.027}{0.1 + r_s} + \frac{0.007}{[0.1 + r_s]^2}. \tag{3.9}$$

Next we describe the long calculation which is required to obtain this simple result. Hubbard (1957) suggested that the longitudinal dielectric function of the homogeneous electron gas can be written usefully in the form

$$\epsilon(q,\omega) = 1 - \frac{\chi(q,\omega)}{1 + G(q)\chi(q,\omega)}, \tag{3.10}$$

$$\chi(q,\omega) = v_q P(q,\omega), \tag{3.11}$$

$$P(q,\omega) = 2\int \frac{d^3k}{(2\pi)^3} \frac{n_{\vec{k}} - n_{\vec{k}+\vec{q}}}{\varepsilon_{\vec{k}} - \varepsilon_{\vec{k}+\vec{q}} - \omega + i\delta}, \tag{3.12}$$

$$G(q) = \frac{1}{2}\frac{q^2}{q^2 + k_F^2}, \tag{3.13}$$

where $\varepsilon_{\vec{k}}$ is the kinetic energy of the electron, and $v_q = 4\pi e^2/q^2$. The factor of $G(q)$ is called a local field correction. It provides a correction to the dielectric function $\epsilon(q,\omega)$ due to exchange and correlation which affects the local environment of the electron. Hubbard's choice (3.13) for $G(q)$ was a first guess. Since then many theorists have tried to find the best form for this function. The best results were obtained by Singwi's group.

Vashishta and Singwi (1972) showed that $G(q)$ can be related to the static structure factor $S(q)$ for the electron liquid which, in turn, is related to a frequency integral over the dielectric function:

$$G(q) = \ -\left(1+\tfrac{2}{3}\tfrac{\partial}{\partial n}\right)\tfrac{1}{n_0}\tfrac{1}{(2\pi)^3}\int d^3q'\vec{q}\cdot\vec{q}'[S(\vec{q}-\vec{q}')-1]/q'^2, \quad (3.14)$$

$$\frac{1}{N} < \rho(\vec{q})\rho(-\vec{q}) > \ = \ N\delta_{\vec{q}=0} + S(q), \quad (3.15)$$

$$n_0 v_q S(q) \ = \ -\int_0^\infty \frac{d\omega}{\pi} Im\left[\frac{1}{\epsilon(q,\omega)}\right]. \quad (3.16)$$

The three equations (3.9), (3.13), and (3.15) link the three functions $G(q)$, $\epsilon(q,\omega)$, and $S(q)$. These three equations are solved self-consistently, which is done for each value of r_s. After obtaining this numerical solution for the local field factor $G(q)$, it is used to calculate the correlation energy. This numerical solution produces a curve which is approximated by the analytical expression in (3.7). Figure 3.1(a) shows a plot of $\varepsilon_c(r_s)$ while Figure 3.1(b) shows $V_c(r_s)$. The Vashishta-Singwi curves are labeled 'VS.' The other curves are explained below. The next method of obtaining the correlation energy is the interpolation scheme of Gordon and Kim (1972). Carr and coworkers (1961, 1964) obtained the exact correlation energy in the limit of high density, and also in the limit of low density,

$$\lim_{r_s\to\infty} \varepsilon_c \ = \ -\frac{0.876}{r_s} + \frac{2.650}{r_s^{3/2}} - \frac{2.94}{r_s^2} - \frac{0.8}{r_s^{5/2}} + O(r_s^{-3})\ldots, \quad (3.17)$$

$$\lim_{r_s\to 0} \varepsilon_c \ = \ 0.0622\ln r_s - 0.096 + 0.018 r_s \ln r_s - 0.02 r_s + \ldots . \quad (3.18)$$

Gordon and Kim (1972) suggested that the correlation energy could be obtained at metallic densities by interpolating between these limiting formulas. Their interpolation was linear for a graph of correlation energy *versus* $\ln r_s$:

$$\varepsilon_c \ = \ -0.12312 + 0.03796\ln r_s, \quad (3.19)$$

$$V_c \ = \ -0.13577 + 0.03796\ln r_s. \quad (3.20)$$

This formula is valid over the range of densities $0.7 < r_s < 10$. The formulas of Gordon and Kim for ε_c and V_c are graphed in Figure 3.1 with the

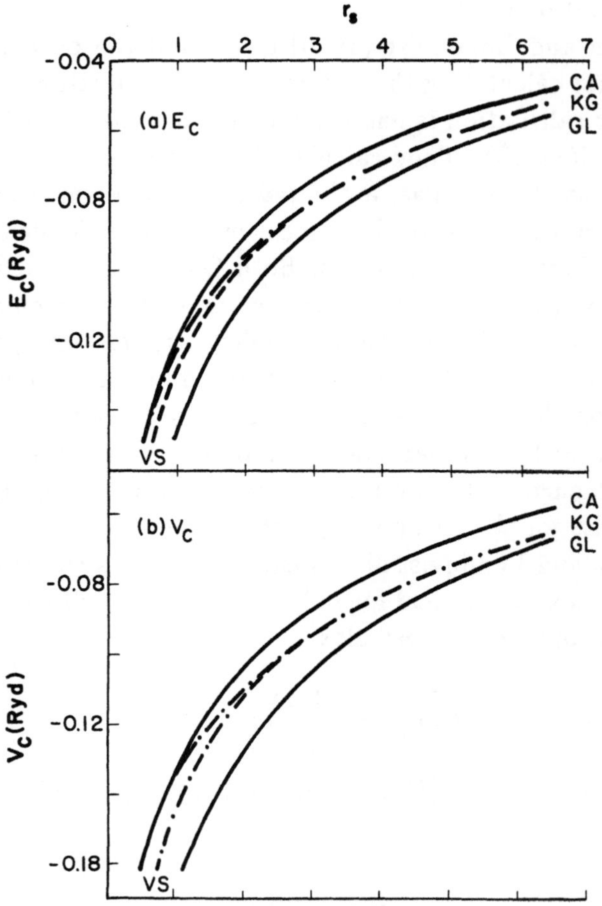

Figure 3.1: The correlation energy (a) and correlation potential (b) as a function of electron density r_s. The top line in each group is Ceperley-Alder while the bottom line is Gunnarsson-Lundqvist. The Gordon-Kim line is shown as dot-dashed and Vashishta-Singwi is dashed. At low values of r_s the KG line coincides with CA, while at higher values of r_s it coincides with VS.

label 'KG'. They are very similar to the results of Vashishta and Singwi. For low density ($r_s > 3$) the two theories agree perfectly. At high density they disagree slightly. This disagreement is a small percentage of the total exchange-correlation energy.

Gunnarsson and Lundqvist (1976) calculated the exchange-correlation potential by first calculating the self-energy of the electron at the Fermi energy k_F. Their calculation is based upon the idea mentioned in Section 2.2 that $V_{xc}(\vec{r}) = \Sigma(k_F(\vec{r}))$. They calculated the self-energy of an electron for the homogeneous electron gas, evaluated it at the Fermi wave vector, and then let the Fermi wave vector be a local function of the density as is done in the Thomas-Fermi approximation. Both Gunnarsson and Lundqvist and Vashishta and Singwi described the properties of the electron gas. Vashishta and Singwi focused their solution on the dielectric response, which is a collective behavior of all of the electrons. Gunnarsson and Lundqvist focused their solution on the properties of the individual electron. Most of the calculations reported by Gunnarsson and Lundqvist were concerned with the spin-dependent parts of the exchange-correlation potential. Here we report only their results for the paramagnetic electron gas.

Gunnarsson and Lundqvist did a long numerical calculation. Afterwards they found a simple analytical formula which fitted well to their numerical output. They proposed the formulas

$$V_c = -0.0666 \ln\left(1 + \frac{11.4}{r_s}\right), \tag{3.21}$$

$$\varepsilon_c = -0.0666 \left[(1 + x^3)\ln(1 + 1/x) + \frac{1}{2}x - x^2 - \frac{1}{3}\right], \tag{3.22}$$

where

$$x = r_s/11.4 \,. \tag{3.23}$$

The correlation potential V_c has the simple formula since that is the quantity they calculated. The ground state energy ε_c is obtained from V_c by integrating (3.2).

Their results are shown in Figure 3.1 as the curves labeled 'GL.' Their values are larger in magnitude than the other theories. We regard the GL theory to be less accurate than the others for describing the correlation energy of the homogeneous electron gas. They overestimate the correlation energy, and provide a correlation potential which is too large in magnitude.

The Gunnarsson-Lundqvist potential energy V_c in (3.21) is widely used in LDA calculations. Although it gives a less accurate ground state energy, it seems to give a better potential. When LDA calculations are done with various choices for V_c, results using GL are better when compared with experimental quantities. Since everyone likes to calculate good results, the GL potential has been adopted widely. This selection is empirical and *ad hoc.* Always remember that the third word in LDA is 'approximation.'

Ceperley and Alder (1980) solved the electron gas on the computer. Classical Monte-Carlo method does not work for electrons since their motion is not described well by Newton's laws of motion. The Monte-Carlo calculation for quantum particles such as electrons was done by assuming a form for the many-particle variational wave function. Monte-Carlo method was used to sample points in phase space. Their calculation was done for systems with a finite number of particles. Their end result is the total ground state energy for the system of N particles. They divide this by N to find the ground state energy per particle. In the range of metallic densities they reported values for $r_s = 1, 2, 5$. The correlation energy is found by subtracting out the contributions from kinetic energy and exchange. The following interpolation formula by Perdew and Zunger (1981) gives the Ceperley-Alder (CA) correlation energy at the three r_s points, and interpolates between them. For $r_s > 1$ then:

$$\varepsilon_C = \frac{\gamma}{1 + \beta_1\sqrt{r_s} + \beta_2 r_s}, \tag{3.24}$$

$$V_c = \frac{\gamma(1 + \frac{7}{6}\beta_1\sqrt{r_s} + \frac{4}{3}\beta_2 r_s)}{(1 + \beta_1\sqrt{r_s} + \beta_2 r_s)^2}, \tag{3.25}$$

where $\gamma = -0.1423$, $\beta_1 = 1.0529$, and $\beta_2 = 0.3334$. A different formula is used at high density, which is important for atomic calculations. For $r_s < 1$ then:

$$\varepsilon_C = B + Dr_s + (A + Cr_s)\ln r_s, \tag{3.26}$$

$$V_c = B - \frac{1}{3}A + \frac{1}{3}(2D - C)r_s + (A + \frac{2}{3}Cr_s)\ln r_s, \tag{3.27}$$

where $A = 0.0311, B = -0.048, C = 0.0020$, and $D = -0.0116$. The correlation potential is found from (3.8). Their results are shown in Figure 3.1 with the label 'CA.' Their curves are smaller in magnitude than the others. Since they performed a variational calculation, their results are an upper bound to the correlation part of the ground state energy. It is reasonable that their curves are higher than those of VS or KG. Presumably a better trial wave function would lower their ground state energy and bring it closer to the other two.

Table 3.1: The Kohn-Sham eigenvalues of atomic neon calculated using correlation by Gunnarsson-Lundqvist, Vashishta-Singwi, Gordon-Kim, and Ceperley-Alder. Values in Rydbergs.

	1s	2s	2p
GL	60.56	2.667	1.021
VS	60.49	2.646	1.002
GK	60.52	2.643	0.999
CA	60.51	2.639	0.995

Our conclusion is that the correlation energy per electron ε_c in the ground state of the homogeneous electron gas is close to the result found by VS and by GK. The variational result of CA is slightly too small in magnitude, while the result of GL is slightly too large.

As an example consider the LDA calculation of the ground state of the neon atom. There are ten electrons in three shells. The radial equation has the form

$$\left[-\frac{\partial^2}{\partial r^2} + \frac{l(l+1)}{r^2} - \frac{20}{r} + V_H(r) + V_{xc}(r) - \lambda_j\right](rR_j) = 0, \tag{3.28}$$

$$n(r) = \frac{1}{2\pi}[R_1^2 + R_2^2 + 3R_3^2]. \tag{3.29}$$

The subscripts $j = 1,2,3$ refer to the three atomic shells. The electron density $n(r)$ is used in the calculation of the two potentials V_H and V_{xc}. The equations are solved self-consistently using techniques to be discussed in Section 3.3. Convergence is attained when the same density $n(r)$ which is used to define the potentials is also found after calculating the eigenfunctions. Table 3.1 shows the Kohn-Sham eigenvalues for the three atomic shells of neon calculated with the four choices for correlation. The eigenvalues for GL correlation are slightly larger. The electrons are more tightly bound using GL correlation. The $2p$ eigenvalue becomes smaller with each case, because the correlation potential is smaller. The Kohn-Sham eigenfunctions obtained using any of these various correlations are quite similar. Graphs of charge densities cannot be distinguished since they differ by less than the width of the ink line.

3.2 Kohn-Sham Eigenfunctions

The numerical techniques for solving the Kohn-Sham eigenfunctions are discussed in this section. These procedures are well known, and are presented in numerous references. For atoms the techniques are closely related to the classic Xα work of Herman and Skillman (1963). For crystalline solids, there are LDA versions of almost all of the traditional band structure methods. The information is provided here for two reasons: (1) it may be unfamiliar to the reader, and (2) the methods for solving for the polarizability are similar and closely related. For the polarizability, one has to solve an inhomogeneous differential equation for the perturbed eigenfunctions $\psi_j^{(\ell)}$. The numerical techniques for solving the inhomogeneous equation are presented in Section 3.3. These techniques are closely related to those for solving for the Kohn-Sham eigenfunctions $\psi_j(\vec{r})$.

The problem at hand is to solve Schrödinger's equation self-consistently to determine the eigenfunction ψ_j and eigenvalue λ_j for each of the occupied atomic orbitals:

$$\begin{aligned} [-\nabla^2 + V(r) - \lambda_j]\psi_j(\vec{r}) &= 0, & (3.30)\\ V(r) &= U(r) + V_H(r) + V_{xc}(r), & (3.31)\\ U(r) &= -\frac{2Z}{r}. & (3.32) \end{aligned}$$

For molecules or crystalline solids, the equations are usually solved in matrix form by expanding $\psi_j(\vec{r})$ in a convenient basis set (e.g., plane waves, Gaussian orbitals, etc.). Here our interest is primarily in closed shell atoms or ions. Then the equations can be solved directly, by following the method of Herman and Skillman (1963).

For an atom with all electrons in filled shells, the potential energy $V(r)$ depends only upon the magnitude of the vector $\vec{r}$. The Coulomb potential from the nucleus is $U = -2Z/r$, where Z is the atomic number. The electron-electron potential V_H and exchange-correlation potential V_{xc} are given in equations (3.6) and, e.g., (3.21). They each depend upon the density $n(r)$ of the electrons which depends upon the eigenfunctions ψ_j. Equation (3.30) must be solved self-consistently. The same electron density used in the potentials must be found at the end after summing over the eigenfunctions of the occupied electron states. This feature of the problem is handled by iteration.

The Hamiltonian in (3.30) is nonrelativistic. The most important relativistic contribution to atomic states is the spin-orbit interaction. Its neglect is an approximation. The present method is unsuitable for atoms of high atomic number, where relativistic effects become important.

The eigenfunctions ψ_j are products of radial functions of r and angular functions of (θ, ϕ). The angular functions are spherical harmonics,

$$\psi_j(\vec{r}) = R_{nl}(r)Y_{lm}(\theta, \phi). \tag{3.33}$$

The angular functions are known. The only problem is to determine the radial function R_{nl}. The principal quantum number is n and the angular momentum is l. In solving for the radial function it is convenient to define $S_{nl}(r) = rR_{nl}(r)$, since it obeys the simple differential equation

$$\left\{-\frac{\partial^2}{\partial r^2} + \frac{l(l+1)}{r^2} + V(r) - \lambda_j\right\} S(r) = 0. \tag{3.34}$$

The subscript labels on $S = rR$ are usually omitted. The function $S(r)$ also has the advantage that it vanishes at the origin for all angular momenta. The Kohn-Sham eigenfunctions are found by solving the radial equation (3.34). When solving the equation for the Kohn-Sham eigenfunction, the eigenvalue λ_j is unknown initially. The eigenvalue is varied until a value is found which satisfies the equation. Searching for the eigenvalue can be tedious, depending on the initial guess, and the equation for S is solved after numerous iterations.

In LDA the interest is in occupied orbitals, which are bound, and all eigenvalues of interest λ_j are negative. For bound states the radial function $S(r)$ has the following boundary conditions: it vanishes at $r \to \infty$ and also at $r \to 0$. We first rewrite equation (3.34) in the form:

$$S''(r) = A_l(r)S(r), \tag{3.35}$$

$$A_l(r) = \frac{l(l+1)}{r^2} + V(r) - \lambda_j. \tag{3.36}$$

The double prime denotes double differentiation. In digital computing, the function is calculated at a discrete set of points r_i. For the moment assume the points are spaced uniformly with a separation η. The subscript i denotes a function evaluated at the point r_i; e.g., $A_i = A(r_i)$. Taylor series are used to expand the function $S(r \pm \eta)$ about the point $S(r)$,

$$S_{i+1} = S_i + \eta S_i' + \frac{\eta^2}{2!}S_i'' + \cdots,$$

$$S_{i-1} = S_i - \eta S_i' + \frac{\eta^2}{2!}S_i'' - \cdots.$$

These two equations are added:

$$S_{i+1} + S_{i-1} = 2S_i + \eta^2 S_i'' + \frac{\eta^4}{12} S_i'''' + O(\eta^6). \tag{3.37}$$

The term S'' can be evaluated using the differential equation (3.35). This produces a three-term difference equation which has an accuracy to $O(\eta^4)$,

$$S_{i+1} = S_i[2 + \eta^2 A_i] - S_{i-1} + O(\eta^4). \tag{3.38}$$

The differential equation has been turned into a three-term recursion relation. The quantity A_i is assumed to be known. If values of S_i are known at two consecutive points, the remaining points can be found by iteration away from these two. Solving the above equation provides an answer with an accuracy of $O(\eta^4)$.

Numerov's method (see, e.g., Atkinson, 1978) is a technique for solving the differential equation to an accuracy of $O(\eta^6)$ while still having a three-term recursion relation. One defines the auxiliary functions

$$\begin{aligned} L(r) &= \left(1 - \frac{\eta^2}{12}\frac{\partial^2}{\partial r^2}\right), \\ T(r) &= L(r)S(r), \\ T_i &= S_i - \frac{\eta^2}{12}S_i'' = S_i[1 - \frac{\eta^2}{12}A_i], \\ S_i &= \frac{T_i}{[1 - \frac{\eta^2}{12}A_i]}. \end{aligned} \tag{3.39}$$

Operate on equation (3.37) by $L(r)$. Generally this operation just changes $LS_i = T_i$. The interesting result occurs in the derivative terms,

$$L\left[\eta^2 S'' + \frac{\eta^4}{12} S''''\right] = \eta^2 S_i'' + O(\eta^8), \tag{3.40}$$

$$= \eta^2 S_i A_i + O(\eta^8), \tag{3.41}$$

$$= \frac{\eta^2 T_i A_i}{1 - \frac{\eta^2}{12}A_i} + O(\eta^8). \tag{3.42}$$

Terms of $O(\eta^4)$ are eliminated. The term S_i'' is evaluated from the differential equation (3.35). The substitution of T_i for S_i is done according to

(3.39). Therefore, we have shown that operating upon equation (3.37) by $L(r)$ produces the following recursion relation:

$$T_{i+1} = T_i \left[2 + \frac{\eta^2 A_i}{1 - \epsilon_i} \right] - T_{i-1} + O(\eta^6), \tag{3.43}$$

$$\epsilon_i \equiv \frac{\eta^2}{12} A_i. \tag{3.44}$$

The three-term recursion relation for T_i is the one which is iterated numerically. At each point the function we need S_i is found easily from (3.39). This equation is just as easy to solve as (3.38), but has the advantage that it is accurate to more decimal places for the same step size.

First we describe the method of solving for the eigenfunctions of the ground state atom. The task is to solve (3.43). Initially, neither the eigenvalue nor the eigenfunction is known, and the two must be found together. For the bound state of an atom the radial function $T(r)$ vanishes at the two end points: $r \to 0$ and also $r \to \infty$.

A simple idea is to guess an initial eigenvalue, and then start iterating outward from the origin. If two successive points are known for T_i, the remaining points can be found by iterating the three-term recursion relation. The first point is $T_0 = 0$. The second point $T_1 = C$ can be selected to be an arbitrary constant. Then iterate the remaining points using (3.43). This solution produces an eigenfunction of the right shape, but of the wrong magnitude. The correct constant C can be chosen later by renormalizing the eigenfunction.

This simple idea does not work. The second-order differential equation (3.34) has two solutions. At large values of r one solution grows exponentially, while the other decays exponentially. The decaying solution is the correct one for bound states. However, numerical rounding errors will always mix in some of the solution which grows exponentially. As r increases in value, this growing solution gets big and dominates the numerical output. Even if the eigenvalue has been chosen accurately, the numerical solution will be incorrect at large values of r.

The correct solution at large r is obtained by starting the iteration at a far point and iterating inward. The solution which grows exponentially during the inward iteration is the one which decays during an outward iteration. It is the solution appropriate for bound states.

The numerical solution for bound states is obtained by the following steps:

1. Choose an initial guess for the eigenvalue λ_j. For atoms or ions, these can be values listed in Herman and Skillman (1963).

2. Choose a match point r_m. This point should be at or near to the maximum value of the eigenfunction. Numerical confusion occurs if this point is near a node.

3. Iterate outward from the origin. The first point is $T_0 = 0$ and the second point is $T_1 = C$ where C is chosen arbitrarily. The iteration is carried outward to the match point. The logarthmic derivative of the eigenfunction is found,
$$L_< = S'_m / S_m. \tag{3.45}$$
Using the logarithmic derivative eliminates the dependence upon the constant C.

4. Iterate inward starting from some far point r_N. For atoms or surfaces the far point is selected well outside the electron density, where the eigenfunction has a value which is exponentially small. In atoms we use values of r_N which are 15-20 atomic units for eigenfunctions for the outer shells of the atoms. One way to choose the first two points is $T_N = 0$ and $T_{N-1} = C'$, where C' is chosen arbitrarily. Another way to choose these points is $T_N = C'$ and $T_{N-1} = C' \exp\{-\eta\sqrt{A_N}\}$. The latter formula is from WKBJ theory, where A_N is found in (3.36). The actual method of choosing these points is not very important. As the inward iteration is started, the desired solution grows exponentially while the wrong one decays exponentially. The answer is independent of the initial conditions after a few iterations. The inward iteration continues until reaching the match point r_m. The logarithmic derivative $L_>$ is found using (3.45) with the eigenfunctions determined from the inward iteration.

5. Choose a new value for the eigenvalue $\lambda_j^{(\text{new})}$,
$$\lambda_j^{(\text{new})} = \lambda_j - \alpha[L_< - L_>].$$
The constant α is found empirically. It is positive and usually less than one.

6. Use the new eigenvalue as λ_j and start again at step 2. Repeat this process until the two logarithmic derivatives $L_<$ and $L_>$ are identical to

within the accuracy of the Numerov procedure. Then the eigenfunction which is a solution to (3.34) has been obtained.

7. Use this procedure to solve for the eigenfunction and eigenvalue for each shell. Afterwards, use them to recalculate the self-consistent density $n(r)$ and the self-consistent potential $V(r)$. Use this new potential to repeat the entire process of finding the eigenfunctions and eigenvalues.

8. Continue all steps until the self-consistent potential is unchanged by the iteration procedure to desired accuracy. Then an LDA solution has been obtained.

The above list describes a straightforward procedure for obtaining a solution to the self-consistent LDA equations for atomic shells. In practice there are tricks which speed up the process of obtaining convergence. In particular, only a fraction of the old potential is replaced by the new potential in the early iterations. It is customary to require convergence in the potential to one part in 10^{-5}, while the eigenvalues converge to one part in 10^8. Table 3.2 gives Kohn-Sham eigenvalues for some closed shell atoms and ions. The positive numbers listed are actually $-\lambda_j$. They were obtained using Gunnarsson-Lundqvist correlation.

The three-term recursion relation (3.43) works for a constant separation η between radial points. It is inconvenient to use only one step size for eigenfunctions of electrons in atoms or ions. The inner shells require a small step size. If this small step is used throughout the atom, too many points are required. Instead, it is easier to increase the step size periodically while iterating outward from the origin. Figure 3.2 shows the region near a change in step size from η to η'. The point r_i is where the step change occurs. The trick of iterating through a change in step size is to recognize that we need to define a new point $r_{(i-1)'}$ which is the point preceding r_i for step size η'. The next point is found using

$$T_{i+1} = T_i \left[2 + \frac{(\eta')^2 A_i}{1 - \frac{(\eta')^2}{12} A_i} \right] - T_{(i-1)'}.$$

A convenient choice is to have η' be an integer multiple of η since then $r_{(i-1)'}$ is a previous point and $T_{(i-1)'}$ is known. Otherwise the value of $T_{(i-1)'}$ is found by interpolation. Then the function $T(r)$ is fitted to a polynomial in

Table 3.2: Negative eigenvalues of the Kohn-Sham eigenfunctions calculated using Gunnarsson-Lundqvist correlation.

Ion	$1s$	$2s$	$2p$	$3s$	$3p$	$3d$	$4s$	$4p$
Ne	60.56	2.667	1.021					
Na^+	75.94	4.720	2.717					
Mg^{++}	93.37	7.314	4.947					
Ar	227.3	21.58	16.91	1.785	0.786			
K^+	256.9	26.11	21.05	3.036	1.861			
Ca^{++}	288.5	31.20	25.73	4.548	3.196			
Cu^+	639.5	76.67	67.53	8.740	5.869	1.060		
Zn^{++}	688.1	84.32	74.75	10.67	7.594	2.341		
Kr	1014.	132.1	120.0	18.59	14.19	6.183	1.656	0.713
Rb^+	1075.	142.5	129.9	21.41	16.78	8.294	2.707	1.625
Sr^{++}	1139.	153.4	140.5	24.52	19.66	10.70	3.935	2.715

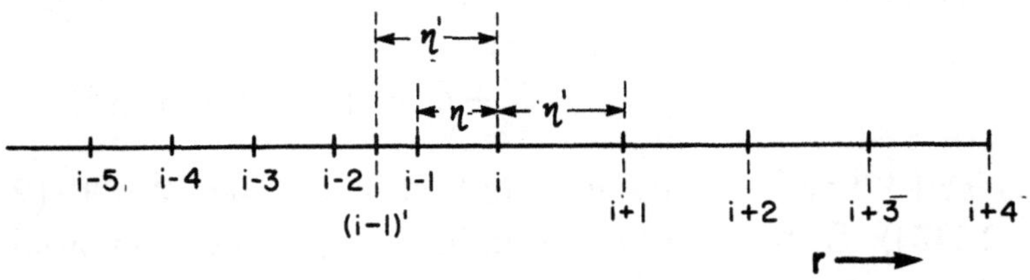

Figure 3.2: The value of $T(r)$ must be found at $r_{(i-1)'}$ when changing step size from η to η'. The change occurs at the point r_i.

the vicinity of r_i:

$$\begin{aligned} T(r) &= T_i + b(r - r_i) + c(r - r_i)^2 + \cdots, \\ T_{(i-1)'} &= T_i - b\eta' + c(\eta')^2 + \cdots. \end{aligned}$$

We use a quartic polynomial which is fitted to five points. This method of changing step size is numerically easy and accurate. The step size can be chosen so that each atomic shell has an adequate number of points to characterize its radial variations. The total number of radial points is usually around 200 to 400.

3.3 Modified Sternheimer Approach

The general formalism of linear and nonlinear response was discussed in Section 2.3. An external potential such as an electric field is applied to an atom or solid. It develops an induced moment, which is caused by the field. The field is regarded as small — even *infinitesimal.* The Kohn-Sham eigenfunctions of the electron in the atom or solid become slightly changed by the applied field. Perturbation theory is used in Section 2.3 to write the general expression for the nth order change $\psi_j^{(n)}(\vec{r})$ in the Kohn-Sham eigenfunction of the jth orbital. In this notation, the orbitals of the ground state are denoted as $\psi_j^{(0)}$. The potentials, eigenfunctions, and eigenvalues of the ground state are often written without the superscript (0). The equation governing the behavior of $\psi_j^{(n)}$ is derived from perturbation theory (see (2.108)),

$$[H_0 - \lambda_j]\psi_j^{(n)}(\vec{r}) = -\sum_{\alpha=1}^{n}[V^{(\alpha)}(\vec{r}) - \lambda_j^{(\alpha)}]\psi_j^{(n-\alpha)}(\vec{r}), \tag{3.46}$$

$$H_0 = -\nabla^2 + U(\vec{r}) + V_H(\vec{r}) + V_{xc}(\vec{r}). \tag{3.47}$$

Formulas for $V^{(\alpha)}$ and $\lambda^{(\alpha)}$ are given in Section 2.3 (see (2.109)–(2.110) and (2.113)–(2.115)). Equation (3.46) is an inhomogeneous differential equation for the unknown function $\psi_j^{(n)}$.

Solving this equation is easy if the right-hand side is a known function of $\vec{r}$. However, the quantities on the right are not known initially. The perturbed Kohn-Sham eigenfunctions $\psi_j^{(\alpha)}$ are assumed known for values of $\alpha < n$. The first step is to solve for the ground state of the system, which determines ψ_j. Then the inhomogeneous equation is solved successively for $n = 1, 2, 3 \ldots$.

The part of the right-hand side which is unknown is $V^{(n)}(\vec{r})$. It depends in (2.113), upon the nth order change in the density $n^{(n)}(\vec{r})$, which itself requires the knowledge of $\psi_j^{(n)}(\vec{r})$. The inhomogeneous differential equation is also nonlinear and nonlocal because $\psi^{(n)}$ occurs in the potential. This part of the problem is usually solved by iteration. In the first step, the potential $V^{(n)}$ is calculated by assuming that $\psi^{(n)} \equiv 0$. The right-hand side of (3.46) is then a known function of $\vec{r}$, which enables this equation to be solved in order to find $\psi_j^{(n)}$. In subsequent iterations, the prior values of $\psi^{(n)}$ are used to calculate $V^{(n)}$. Typically, 8-10 iterations produce convergence in the polarizability to better than 1%. The right-hand side of (3.46) is abbreviated in what follows by the symbol $u(\vec{r})$. It is regarded as a known function while we are trying to solve the equation

$$[H_0 - \lambda_j]\psi_j^{(n)} = u(\vec{r}). \tag{3.48}$$

There are two different numerical techniques which have been used to solve this inhomogeneous differential equation. The first one is called the *modified Sternheimer method.* Sternheimer (1954) developed this method while solving for the dipole and quadrupole response of atoms. His early work did not employ an exchange-correlation potential, nor was the potential $V^{(n)}$ found self-consistently. The right-hand side of his (3.46) contained only the external potential. That is why the phrase "modified" is used to describe the present method. Nevertheless, Sternheimer pioneered the method of solving an inhomogeneous differential equation directly in order to obtain the response functions. The modified Sternheimer method solves this equation as a differential equation. The method of solution is described in the present section. It is numerically fast, easy, and straightforward. The other technique of solving this differential equation is called the *Green's function method.* It is described in the next section.

Once again, we restrict our attention to a spherically symmetric geometry in the ground state. Spherical coordinates are used for atoms with closed atomic shells. The factors depending upon angles are derived analytically. The dependence upon the radial variable r is solved numerically. This method is now described.

First consider the terms which depend upon the angular factors. The technique is best explained with a simple example. Consider the response of an atom to an external potential of the form

$$V_M(\vec{r}) = \mathcal{E}_M(r)P_M(\nu), \tag{3.49}$$

where $\mathcal{E}_M$ is an infinitesimal field, and P_M is a Legendre polynomial of index M and argument $\nu = \cos(\theta)$. For dipole response $M = 1$, for quadrupole response $M = 2$, etc. When calculating the first-order response, $n = 1$ and $u(\vec{r}) = -[V^{(1)}_{SCF,M}(\vec{r}) - \lambda^{(1)}_j]\psi_j(\vec{r})$. The relationship between the applied potential and that of the self-consistent field is (see Section 2.3.4):

$$V^{(1)}_{SCF,M}(\vec{r}) = V_M(\vec{r}) + 2\int d^3r' \frac{n^{(1)}(\vec{r}')}{|\vec{r}-\vec{r}'|} + n^{(1)}(\vec{r})\frac{\partial V_{xc}}{\partial n_0}, \quad (3.50)$$

$$\lambda^{(1)}_j = \int d^3r V^{(1)}_{SCF,M}(\vec{r})|\psi_j(\vec{r})|^2. \quad (3.51)$$

There is a theorem (Eaves and Epstein, 1974) which makes the calculation much easier. It states that, for an atom with all electrons in filled shells, the change in density $n^{(1)}$ due to the perturbation V_M has its angular dependence given entirely by P_M. The potential of the self-consistent field is then determined by the same angular dependence:

$$n^{(1)}(\vec{r}) = n^{(1)}_M(r)P_M(\nu), \quad (3.52)$$

$$V^{(1)}_{SCF,M}(\vec{r}) = v^{(1)}_{SCF,M}(r)P_M(\nu), \quad (3.53)$$

$$v^{(1)}_{SCF,M}(r) = \mathcal{E}_M(r) + \frac{8\pi}{2M+1}\int dr' \frac{r^M_<}{r^{M+1}_>} n^{(1)}_M(r') + n^{(1)}_M(r)\frac{\partial V_{xc}}{\partial n_0}, \quad (3.54)$$

where $r_<$ and $r_>$ are the smaller and greater of (r, r'). For atoms with partially filled p- or d-shells the angular response is much more complicated. These cases have not yet been considered.

The label j on the atomic eigenstate denotes (nlm); n is the radial quantum number and (lm) are the angular quantum numbers. The angular dependence of ψ_j is given by the spherical harmonic $Y_{lm}(\theta, \phi)$. The function $u(\vec{r})$ on the right-hand side of the inhomogeneous equation (3.48) has its angular dependence determined by the product of angular functions:

$$P_M(\nu)Y_{lm}(\theta,\phi) = \sum_{l'=|l-M|}^{l+M} C(M0; lm|l', m)Y_{l',m}(\theta,\phi).$$

The quantity $C(M0; lm|l'm)$ is a Gaunt coefficient (see Sakurai, 1985). The function $u(\vec{r})$ is also expanded as a summation over l'. The radial coefficient is defined as $c_{l'}(r)$:

$$u(\vec{r}) = \sum_{l'=|l-M|}^{l+M} C(M0;lm|l'm)c_{l'}(r)Y_{l'm}(\theta,\phi), \tag{3.55}$$

$$c_{l'}(r) = -[v^{(1)}_{SCF,M}(r) - \lambda^{(1)}_{nl}\delta_{ll'}]R_{nl'}(r). \tag{3.56}$$

The eigenvalue term arises only when $l' = l$.

So far the discussion has been for the first-order perturbation where $n = 1$. The same type of terms are found for $n > 1$. For these cases $u(\vec{r})$ is a summation of terms $[V^{(\alpha)} - \lambda^{(\alpha)}]\psi^{(n-\alpha)}$. Each term has angular factors similar to those discussed above, and contributes to the coefficient $c_{l'}(r)$.

The first-order eigenfunction $\psi_j^{(1)}$ has the same angular dependence as the function $u(\vec{r})$. The perturbed eigenfunction can be written exactly as

$$\psi_j^{(1)}(\vec{r}) = \sum_{l'=|l-M|}^{l+M} \sum_m C(M0;lm|l'm)R^{(1)}_{nl'}(r)Y_{l'm}(\theta,\phi). \tag{3.57}$$

The quantity which is most useful is the change in density $n^{(1)}$. By collecting the various results, its radial part can be written as

$$n^{(1)}_M(r) = \sum_{nll'} R_{nl}(r)R^{(1)}_{nl'}(r)D(l,l';M), \tag{3.58}$$

$$D(l,l';M)P_M(\nu) = 2\sum_m C(M0;lm|l'm)Y^*_{lm}(\theta,\phi)Y_{l'm}(\theta,\phi). \tag{3.59}$$

The coefficients $D(l,l';M)$ are the only angular information needed to obtain the perturbed density $n^{(1)}_M(r)$ for an atom with all electrons in closed shells. Some values of the quantity are shown in Table 3.3. These coefficients were first derived by Sternheimer for s- and p-states. Note that parity is conserved. Coefficients which do not conserve parity are zero. For numerical computing it is convenient to have a formula instead of a table of numbers. The formulas for $D(l,l';M)$ for values $l = 0, 1, 2$ are listed below.

$$D(0,M;M) = \frac{1}{\pi}$$

$$D(1,M+1;M) = \frac{3}{\pi}\frac{M+1}{2M+1}$$

$$D(1,M-1;M) = \frac{3}{\pi}\frac{M}{2M+1}$$

Table 3.3: The coefficients $D(l, l'; M)$.

l	l'	M=1	M=2	M=3
0	1	$1/\pi$	0	0
0	2	0	$1/\pi$	0
0	3	0	0	$1/\pi$
1	0	$1/\pi$	0	0
1	1	0	$6/5\pi$	0
1	2	$2/\pi$	0	$9/7\pi$
1	3	0	$9/5\pi$	0
1	4	0	0	$12/7\pi$
2	0	0	$1/\pi$	0
2	1	$2/\pi$	0	$9/7\pi$
2	2	0	$10/7\pi$	0
2	3	$3/\pi$	0	$4/3\pi$
2	4	0	$18/7\pi$	0
2	5	0	0	$50/21\pi$

$$\begin{aligned} D(2, M-2; M) &= \frac{15}{2\pi}\frac{M(M-1)}{(2M-1)(2M+1)} \\ D(2, M; M) &= \frac{5}{\pi}\frac{M(M+1)}{(2M-1)(2M+3)} \\ D(2, M+2; M) &= \frac{15}{2\pi}\frac{(M+1)(M+2)}{(2M+1)(2M+3)}. \end{aligned}$$

The response function of the atom to the external field is

$$p_M = \int d^3r V_M(\vec{r}) n^{(1)}(\vec{r}), \tag{3.60}$$

$$= -\frac{4\pi}{2M+1}\int r^2 dr \mathcal{E}_M(r) n_M^{(1)}(r). \tag{3.61}$$

In deriving the second equation we have used the fact that both V_M and $n^{(1)}$ have an angular factor of $P_M(\nu)$. The angular integral gives the prefactor of $4\pi/(2M+1)$.

When calculating the dipole polarizability of an atom or ion, the external field is

$$V_1 = -e\vec{\mathcal{E}} \cdot \vec{r} = -e\mathcal{E}r\nu \tag{3.62}$$

when the applied field is in the z-direction. Since $n^{(1)} \sim V_1$, then $n^{(1)} \sim e$ and the dipole polarizability has a factor of $e^2 = 2$ in atomic units. So in calculating the polarizability, we usually set $\mathcal{E}_1 = r$ and account for the factor of e^2 by multiplying the final result by an additional factor of 2.

So far the discussion has been about the response of an atom to an external perturbation whose angular dependence is $P_M(\nu)$. One can show that the same linear response function is found for an external perturbation with an angular dependence of $P_M(\nu)\exp(im\phi)$. The additional factor of $\exp(im\phi)$ does not change the polarizability. An external potential $V_{ext}(\vec{r})$ can be decomposed into a series over spherical harmonics,

$$V_{ext}(\vec{r}) = \sum_{lm} v_{ext}(lm, r) Y_{lm}(\theta, \phi). \tag{3.63}$$

The linear response can be calculated for each term of this series separately. The total linear response is the summation of all of these contributions. In this way the linear response can be calculated by the present methods for any external potential $V_{ext}(\vec{r})$ which is applied to the atom. This completes our discussion of the angular factors in the calculation of polarizabilities.

The solution to the inhomogeneous differential equation (3.48) contains angular functions and a radial function $R_j^{(n)}(r)$. This latter function is found by solving the radial part of the inhomogeneous equation. It is convenient to define

$$X(r) = rR_j^{(n)}(r), \tag{3.64}$$

$$\left\{-\frac{\partial^2}{\partial r^2} + \frac{l'(l'+1)}{r^2} + V(r) - \lambda_j\right\} X(r) = rc(r), \tag{3.65}$$

$$V(r) = U(r) + V_H(r) + V_{xc}(r). \tag{3.66}$$

The atomic eigenfunction $S = rR$ obeys a differential equation similar to (3.34). These functions are assumed to be known: the first step in finding the response functions is to find the Kohn-Sham eigenfunctions of the ground state.

The equation for X is much easier to solve than the eigenvalue equation for $S(r)$. In solving for $X(r)$ it is assumed that $c(r)$ is known, so that the only unknown is $X(r)$. $X(r)$ obeys the same boundary conditions as $S(r)$: $X(r)$ vanishes at $r \to \infty$ and also at $r \to 0$. Since $X(r)$ has the appearance of an eigenfunction, it may be found by a numerical procedure which is a slight variation on the method of finding the radial function $S(r)$ described earlier. This method is now described.

Rewrite the equation (3.65) in the form

$$X''(r) = A_l(r)X(r) + B(r), \tag{3.67}$$

$$A_l(r) = \frac{l(l+1)}{r^2} + V(r) - \lambda_j, \tag{3.68}$$

$$B(r) = rc(r). \tag{3.69}$$

The double prime denotes double differentiation. Again, assume the function is calculated at a set of points r_i which are equally spaced with a separation η. The subscript i denotes a function evaluated at the point r_i; e.g., $B_i = B(r_i)$. Taylor series are used to expand the function $X(r \pm \eta)$ about the point $X(r)$. Following the steps in Section 3.2 one can derive the relation

$$X_{i+1} + X_{i-1} = 2X_i + \eta^2 X_i'' + \frac{\eta^4}{12} X_i'''' + O(\eta^6). \tag{3.70}$$

This expression can be turned into a three-term recursion relation by using (3.67) to evaluate X'' and by neglecting terms of $O(\eta^4)$. A more accurate recursion relation is obtain by using Numerov's method. To this end we define the auxiliary function $Y(r) = L(r)X(r)$, where

$$L(r) = \left(1 - \frac{\eta^2}{12}\frac{\partial^2}{\partial r^2}\right), \tag{3.71}$$

$$Y_i = X_i - \frac{\eta^2}{12} X_i'' = X_i \left[1 - \frac{\eta^2}{12} A_i\right] - \frac{\eta^2}{12} B_i, \tag{3.72}$$

$$X_i = \frac{Y_i + \frac{\eta^2}{12} B_i}{1 - \frac{\eta^2}{12} A_i}. \tag{3.73}$$

The three term recursion relation for Y_i is obtained by operating on equation (3.70) by $L(r)$. This eliminates terms of $O(\eta^4)$ and yields an equation for Y_i which is analogous to (3.42) for T_i:

$$Y_{i+1} = Y_i \left[2 + \frac{\eta^2 A_i}{1 - \epsilon_i}\right] - Y_{i-1} + \frac{\eta^2}{1 - \epsilon_i} B_i + O(\eta^6), \tag{3.74}$$

$$\epsilon_i \equiv \frac{\eta^2}{12} A_i. \tag{3.75}$$

The three-term recursion relation for Y_i is the one which is iterated numerically. At each point the function X_i is found easily from (3.73). The process of solving (3.74) for Y_i contains many steps similar to solving for the Kohn-Sham eigenfunctions. The process is much faster since there is no search for an eigenvalue. Our first calculations used a method which obtained an answer after a half-dozen iterations. Our current procedure, suggested by Senatore and Subbaswamy, obtains the same answer after only two steps without any iterations. This latter method is now described.

The basic steps in solving (3.74) are the following:

1. Choose a match point. For best numerical accuracy, this should be the same point chosen for the solution of the unperturbed eigenfunction.
2. Iterate outward from the origin to the match point.
3. Iterate inward from the far point r_N to the match point.
4. Compare the inward and outward solutions at the match point.

The features which must be described are: (i) how to start the iterations at $r = 0$ and at r_N, and (ii) how to compare them at the match point.

The Kohn-Sham eigenfunction obeys the equation $S'' = A_l S$ and the change in the eigenfunction obeys the equation $X'' = A_{l'} X + B$. The solution for $X(r)$ depends upon whether $A_{l'} = A_l$, i.e., whether $l = l'$. This point is quite important.

When $l' = l$ the function $X(r)$ has an arbitrary admixture of $S(r)$ in its solution. If $X(r)$ is a solution to (3.65) then $X(r) + \epsilon S(r)$ is also a solution where ϵ is arbitrary. The homogeneous equation is a solution to the inhomogeneous one. This situation occurs, for example, when calculating the quadrupole polarizability of an atom: when finding $\psi^{(1)}$ for p- or d-states for $M = 2$. It also occurs when calculating nonlinear response functions.

The two cases of $l' = l$ and $l' \neq l$ must be discussed separately. The latter is easier, and is done first.

[$l' \neq l$]: The notation $X_{<,>}$ is used to denote the function at the match point as determined by outward $X_<$ and inward $X_>$ iterations. A prime denotes the derivative. When $l' \neq l$, the two conditions at the match point are

$$X_< = X_>, \tag{3.76}$$
$$X'_< = X'_>. \tag{3.77}$$

The function for $r < r_m$ is found by defining two auxiliary functions which are called $g(r)$ and $f(r)$. The steps to determine $X_<$ are:

1. Define $X = ag + f$ where a is a constant which needs to be determined.

2. Determine $g(r)$ by solving the equation $g'' = Ag$ outward from the origin with the initial conditions $g_0 = 0, g_1 = 1$. The function $A(r)$ is the same one in the equation for $X'' = AX + B$. The equation $g'' = Ag$ is solved by using Numerov's method of constructing a three-term recursion relation such as (3.42). It is integrated outward from the origin to the match point. This procedure determines $g_<$ and $g'_<$.

3. Determine $f(r)$ by solving the equation $f'' = Af + B$ outward from the origin with the initial conditions that $f_0 = 0, f_1 = 0$. The inhomogeneous equation is solved using a recursion relation similar to (3.74). Nonzero values are found for f_i because of the inhomogeneous term. This procedure determines $f_<$ and $f'_<$.

An identical procedure is used to find the solution for $r > r_m$.

1. Define $X = bg + f$ where b is a constant to be determined below.

2. Determine g by solving $g'' = Ag$ with the boundary conditions at the far point r_N of $g_N = 0, g_{N-1} = 1$. Use the recursion relation to iterate inward to the match point. This procedure determines $g_>$ and $g'_>$.

3. Determine f by solving $f'' = Af + B$ with the boundary conditions at the far point of $f_N = 0, f_{N-1} = 0$. Iterate inward to the match point. This procedure determines $f_>$ and $f'_>$.

These steps produce values for $X_<$, $X_>$, and their derivatives. Equations (3.76) and (3.77) can be written as

$$ag_< - bg_> = -f_< + f_>, \tag{3.78}$$

$$ag'_< - bg'_> = -f'_< + f'_>. \tag{3.79}$$

Only a and b are unknown in these equations. The quadratic equations are solved easily to determine these two constants. Then the solution for $X(r)$ is obtained for all values of r. This method of solution is fast. The equations are only solved twice: once to find g and once to find f. Then one solves (3.78) and (3.79), and the solution is determined completely.

[$l' = l$]: The procedure described above for $l' \neq l$ does not work when $l' = l$. The problem stems from the fact that the matching conditions of X and of X' are no longer linearly independent. One can show that

$$X'_< - X'_> = L[X_< - X_>],$$

where L is the logarithmic derivative (3.45) at the match point of the eigenfunction of the Kohn-Sham equation. This equation is only valid when the inhomogeneous equation $X'' = A'X + B$ and the homogeneous equation $S'' = AS$ have the same function $A = A'$ which occurs only when $l' = l$. One condition at the match point is that $X_< = X_>$. The above result shows that matching derivatives provides no new information. Two matching conditions are still required, so another matching condition must be found.

Otherwise, the basic procedure is the same as before. For $r < r_m$ we define $X = ag + f$ where a is an unknown constant and f and g are determined exactly as before. For $r > r_m$ we define $X = bg + f$ where b is an unknown constant and f and g are determined by inward iteration from the far point. In this case $g(r)$ is proportional to the eigenfunction $S(r)$. Two equations at the match point are required to determine the two constants a and b.

If $X(r)$ is a solution of the inhomogeneous equation, then so is $X(r) + \epsilon S(r)$, where $S(r)$ is a solution of the homogeneous equation. How is ϵ determined? Since the number of electrons is unchanged by the external fields, the perturbed densities obey the constraint

$$\int d^3 r n^{(k)}(\vec{r}) = 0 \tag{3.80}$$

for all $k > 0$. Actual expressions for $n^{(k)}$ are given in (2.105)–(2.106). The condition (3.80) can be written as

$$\int_0^\infty dr X(r) S(r) = K, \tag{3.81}$$

where K is a known constant. The constant is zero for first-order perturbations. This constraint can be utilized as the additional matching condition. Since $S(r)$ is known from the solution of the Kohn-Sham equations, the above expression can be rewritten as

$$\int_0^{r_m} dr X_<(r) S(r) + \int_{r_m}^{r_N} dr X_>(r) S(r) = K, \tag{3.82}$$

or

$$\begin{aligned} a \int_0^{r_m} dr g_<(r) S(r) + b \int_{r_m}^{r_N} dr g_>(r) S(r) \\ = K - \int_0^{r_m} f_<(r) S(r) - \int_{r_m}^{r_N} f_>(r) S(r). \end{aligned} \tag{3.83}$$

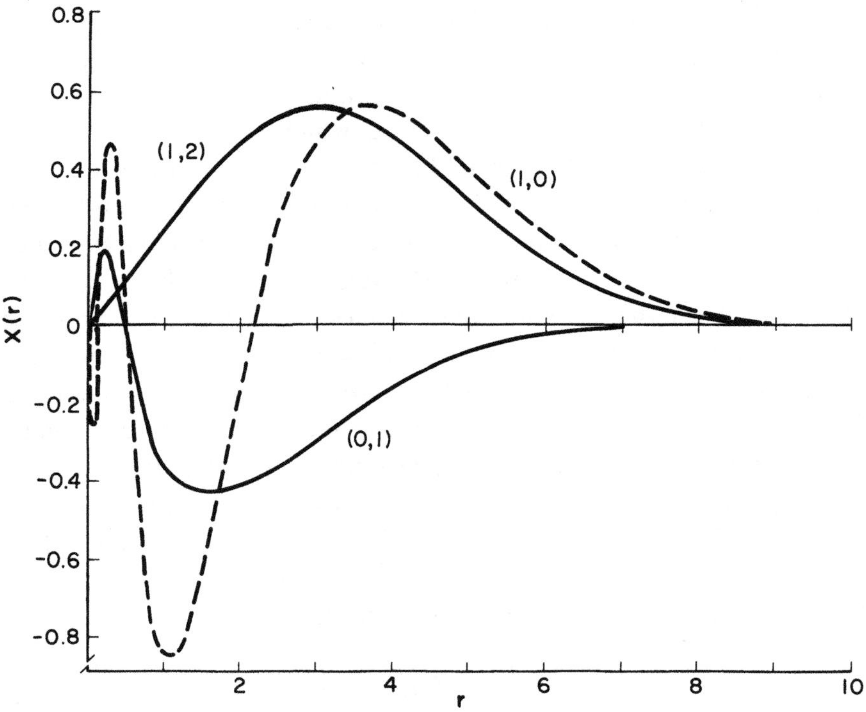

Figure 3.3: $X(r)$ for an argon atom in an electric field. The curve labeled (0,1) is the change in the $3s$ state, while the curves labeled (1,0) and (1,2) are changes in the $3p$ state. The labels (l, l') denote the angular momentum of the atomic state, and of $X(r)$, respectively.

Equation (3.78) is combined with the above expression to provide two equations for the constants a and b. This method determines $X(r)$ everywhere, and solves (3.65) for the cases where $l' = l$.

Figure 3.3 shows the function $X(r)$ for an argon atom in an electric field. The curves are labeled (l, l') where l is the angular momentum of the atomic state, while l' is the angular momentum of $X(r)$. The curve labeled $(0, 1)$ is the change in the $3s$ state due to the field. The curves labeled $(1, 0)$ and $(1, 2)$ are the changes in the $3p$ atomic state. The curve $(1, 0)$ crosses zero three times, while $(1, 2)$ has no crossing. Both are orthogonal to the $3p$ radial function $R_{31}(r)$ which crosses zero once.

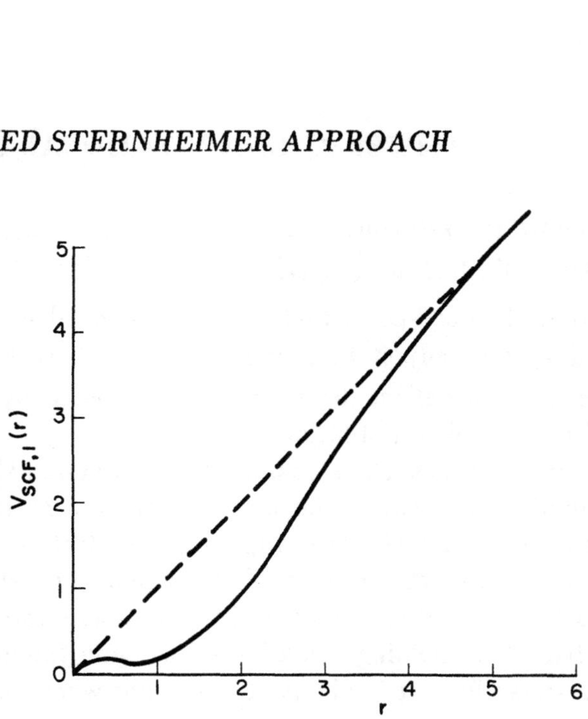

Figure 3.4: The self-consistent potential $v^{(1)}_{SCF,M}(r)$ for an argon atom in an electric field ($M = 1$). The dashed line gives the radial potential r which is the unscreened potential. The solid line is the screened potential $v^{(1)}_{SCF,M}$.

The function $c(r)$ in (3.65) contains the self-consistent potential $v^{(1)}_{SCF,M}(r)$ in (3.53). That potential depends upon $n^{(1)}_M(r)$, which in turn depends upon $X(r)$. After solving for $X(r)$, it is used to calculate a new value of $n^{(1)}$, of $v^{(1)}_{SCF,M}$, and of $c(r)$. Then one solves again for $X(r)$ using this new potential. This process is iterated until the function $X(r)$ is unchanged by further iterations.

The prefix *Time-Dependent* is used to denote the process of calculating the potential $v^{(1)}_{SCF,M}$ until it is self-consistent. This method of calculating the linear polarizability is called the *Time-Dependent Local Density Approximation* or TDLDA. If we had used $v^{(1)}_{SCF,M} = \mathcal{E}_M$, instead of (3.55), the potential would not be found self-consistently. Such a calculation is called an 'LDA' calculation of the polarizability. The prefix 'TD' is omitted when self-consistency is not attempted for the perturbed quantities. Similarly the Hartree-Fock calculations of polarizability are denoted by 'TDHFA' or 'HFA' depending upon whether self-consistency is imposed on the applied potential.

Figure 3.4 shows the self-consistent potential for an argon atom in an electric field. The solid line is $v_{SCF,M}^{(1)}(r)$, while the dashed line is the unscreened potential. The self-consistent applied potential $v_{SCF,M}^{(1)}$ describes the screening. Electrons adjust their motion to try to screen the applied fields. In the present formalism the screening is included by using $v_{SCF,M}^{(1)}$ instead of $\mathcal{E}_M$ for the applied potential.

For a conducting system such as a metal, the conduction electrons screen the applied potential and reduce the field to nearly zero in the interior. An atom is not a metal, yet the screening is quite effective. The screened potential is quite small in the interior of the atom. The electrons in the outer shells adjust their motion to reduce the field acting upon the electrons in the inner shells. This shielding has a big effect upon the numerical value of the dipole polarizability. It is about 40% smaller when calculated using TDLDA compared to LDA.

The screening is far less important for applied fields with values of angular momentum $M > 1$. These applied fields have a radial dependence given by r^M. This factor is small in the interior of the atom. There is no need for screening to make it smaller. The screening is most important for perturbations with $M = 0$ or with $M = 1$. For these two cases at least it is important to use TDLDA instead of LDA in calculating the response functions.

3.4 Green's Function Method

In Section 3.3 the modified Sternheimer method was described which circumvents the sum over intermediate states encountered in perturbation methods by directly solving the inhomogeneous Schrödinger equation. An alternate, but entirely equivalent method is to solve the inhomogeneous equation by means of the Green's function of the differential operator. This formulation of LDA response, originally used by Zangwill and Soven (1980) and Stott and Zaremba (1980), has been widely applied in the study of surfaces, small metal particles, etc. For *nonlinear* response and *frequency-dependent* response this formulation turns out to be somewhat more convenient computationally than the modified Sternheimer approach.

We begin with the simple case of static, first order response. Writing out (3.48) explicitly,

$$[H_0 - \lambda_j]\psi_j^{(1)}(\vec{r}) = -V^{(1)}(\vec{r})\psi_j^{(0)}(\vec{r}) \equiv u(\vec{r}), \tag{3.84}$$

one formally takes the inverse of the operator on the left-hand side of the equation by inserting a complete set of eigenstates of the ground state Hamiltonian H_0. Thus, one gets

$$\psi_j^{(1)}(\vec{r}) = \int d^3r' G_j(\vec{r},\vec{r}\,')u(\vec{r}\,'), \tag{3.85}$$

$$G_j(\vec{r},\vec{r}\,') = \sum_{l\neq j} \frac{\psi_l(\vec{r})\psi_l^*(\vec{r}\,')}{\lambda_l - \lambda_j}, \tag{3.86}$$

where we have introduced the Green's function $G_j(\vec{r},\vec{r}\,')$ for the jth ground state orbital.

From the expression for $n^{(1)}$ in (2.84), we get for the induced density response

$$n^{(1)}(\vec{r}) = \int d^3r\,'\chi^{(1)}(\vec{r},\vec{r}\,')V^{(1)}(\vec{r}\,'), \tag{3.87}$$

where

$$\chi^{(1)}(\vec{r},\vec{r}\,') = \sum_{j(occ.)} \{\psi_j^*(\vec{r})G_j(\vec{r},\vec{r}\,')\psi_j(\vec{r}\,') + \text{c.c.}\}. \tag{3.88}$$

Recall that the ψ_j are the solutions of the ground state Kohn-Sham equations. Thus, the problem is reduced to that of finding the Green's function. But this involves the occupied as well as all the unoccupied orbitals, as can be seen from the sum in (3.86). However, the Green's function obeys the inhomogeneous equation

$$[H_0 - \lambda_j]G_j(\vec{r},\vec{r}\,') = \delta(\vec{r} - \vec{r}\,'), \tag{3.89}$$

and may be solved directly by standard techniques. Before discussing the details, it is convenient to discuss the general frequency-dependent case. General formulas for frequency-dependent nonlinear response functions in a form useful for LDA calculations were given by Zangwill (1983) and Senatore and Subbaswamy (1987). The kth order density response at frequency ω to a perturbation

$$V(t) = \int d^3r\; n(\vec{r},t)V_{ext}(\vec{r},t) \tag{3.90}$$

may be written quite generally as

$$\begin{aligned}\delta n(\vec{r},\omega) = {} & \lim_{\eta\to 0+} \sum_{k=1}^{\infty} \int d^3r_1 \cdots \int d^3r_k \int \frac{d\omega}{2\pi} \\ & \chi^{(k)}(\vec{r},\vec{r}_1,\ldots,\vec{r}_k;\omega_1 + i\eta,\ldots,\omega_k + i\eta) \\ & \times V(\vec{r}_1,\omega_1)\cdots V(\vec{r}_k,\omega_k)2\pi\delta(\omega - \omega_1 - \cdots - \omega_k).\end{aligned} \tag{3.91}$$

The infinitesimal η guarantees the causality of the response. The potential $V(\vec{r},\omega)$ entering the integrals above is to be regarded as the effective potential, i.e., the sum of the external and the induced potentials. Note that even in the case of a monochromatic external potential, the screened potential can have other frequency components because of nonlinear response. We give below some explicit expressions for first- and second-order responses. The method of their derivation is exactly as in the simple, static case considered above.

The single-particle retarded Green's function $G^R(\vec{r},\vec{r}\,';\omega)$ is introduced as

$$G^R(\vec{r},\vec{r}\,';\omega) = \sum_i \frac{\psi_i^*(\vec{r})\psi_i(\vec{r}\,')}{\omega - \lambda_i + i\eta}, \tag{3.92}$$

where the sum runs over all independent values of the one-particle eigenvalues. The first two response functions are given by

$$\begin{aligned}\chi^{(1)}(\vec{r},\vec{r}_1;\omega_1) &= \sum_{j(occ.)} \{\psi_j^*(\vec{r})\psi_j(\vec{r}_1)G^R(\vec{r},\vec{r}_1;\lambda_j+\omega_1) \\ &\quad +\psi_j(\vec{r})\psi_j^*(\vec{r}_1)G^{R^*}(\vec{r},\vec{r}_1;\lambda_j-\omega_1)\},\end{aligned} \tag{3.93}$$

$$\chi^{(2)}(\vec{r},\vec{r}_1,\vec{r}_2;\omega_1,\omega_2) \tag{3.94}$$

$$\begin{aligned}&= \frac{1}{2}\sum_{j(occ.)} \{\psi_j^*(\vec{r})\psi_j(\vec{r}_2)G^R(\vec{r},\vec{r}_1;\lambda_j+\omega_1+\omega_2)G^R(\vec{r}_1,\vec{r}_2;\lambda_j+\omega_2) \\ &\quad +\psi_j(\vec{r})\psi_j^*(\vec{r}_2)G^{R^*}(\vec{r},\vec{r}_1;\lambda_j-\omega_1-\omega_2)G^{R^*}(\vec{r}_1,\vec{r}_2;\lambda_j-\omega_2) \\ &\quad +\psi_j(\vec{r})\psi_j^*(\vec{r}_2)G^R(\vec{r},\vec{r}_1;\lambda_j+\omega_1)G^{R^*}(\vec{r},\vec{r}_2;\lambda_j-\omega_2) \\ &\quad + [\text{permutations of } (\vec{r}_1,\omega_1;\vec{r}_2,\omega_2)]\}\end{aligned} \tag{3.95}$$

As mentioned already, the Green's function involves all occupied and unoccupied Kohn-Sham orbitals, and hence, cannot be determined practically via this route. It obeys the inhomogeneous Schrödinger equation

$$[H_0 - \omega]G_j^R(\vec{r},\vec{r}\,';\omega) = \delta(\vec{r}-\vec{r}\,'). \tag{3.96}$$

For the case of a spherically symmetric ground state potential, which is relevant to closed shell atoms and ions, the equation is solved by first performing the spherical harmonic expansion,

$$G^R(\vec{r},\vec{r}\,';\omega) = \sum_{\ell,m} G_\ell^R(r,r';\omega)Y_\ell^m(\hat{r})Y_\ell^{m^*}(\hat{r}\,'). \tag{3.97}$$

The radial part is constructed by the well-known prescription [see, e.g., Arfken (1982)]

$$G_\ell^R(r, r'; \omega) = \frac{u_\ell^{(1)}(r_<) u_\ell^{(2)}(r_>)}{r r' W[u_\ell^{(1)}, u_\ell^{(2)}]}, \tag{3.98}$$

where $r_<$ and $r_>$ denote, respectively, the smaller or greater of (r, r'). Here $u_\ell^{(1)}$ and $u_\ell^{(2)}$ are solutions of

$$\left[-\frac{d^2}{dr^2} + \frac{\ell(\ell+1)}{r^2} + U(r) + V_H(r) + V_{xc}(r) - \omega\right] u_\ell^{(i)}(r) = 0 \tag{3.99}$$

which are regular at the origin and at infinity, respectively. $W[u_\ell^{(1)}, u_\ell^{(2)}]$ denotes the Wronskian:

$$W[f, g] = f\frac{dg}{dr} - \frac{df}{dr}g. \tag{3.100}$$

The functions $u_\ell^{(1)}$ and $u_\ell^{(2)}$ needed for the construction of the Green's functions are obtained by solving the inhomogeneous Schrödinger equation, (3.99) using the Numerov procedure exactly as described in the previous section. $u^{(1)}$ is obtained by imposing the boundary condition $u^{(1)}(0) = 0$, while $u^{(2)}$ is obtained by imposing $u^{(2)}(\infty) = 0$. The slopes are arbitrary because of the presence of the Wronskian in the denominator. Note that if $\hbar\omega$ is greater than the ionization threshold, the boundary condition at ∞ is modified so as to correspond to an outgoing wave.

The frequency-dependent screened potential that enters the calculation of the induced response, (3.91), can be deduced from the discussion in Section 2.3 [cf. (2.111)–(2.112)]:

$$V^{(\alpha)}(\vec{r}, \omega) = V_{ext}(\vec{r}, \omega) + 2\int d^3r' \frac{n^{(\alpha)}(\vec{r}', \omega)}{|r - r'|} + \delta V_{xc}^{(\alpha)}(\vec{r}, \omega). \tag{3.101}$$

It is not useful to write down general expressions for $\delta V_{xc}^{(\alpha)}$. For the case of a monochromatic external potential of frequency ω the first-order response will be at ω, while the second-order response can be at 2ω as well as at zero frequency as already discussed in Section 2.3. We give below two typical expressions:

$$V_{xc}^{(1)}(\vec{r}, \omega) = \left[\left(\frac{\partial V_{xc}}{\partial n}\right)_{\substack{n = n_0(\vec{r}) \\ \omega = 0}}\right] n^{(1)}(\vec{r}, \omega), \tag{3.102}$$

$$V_{xc}^{(2)}(\vec{r}, 2\omega) = \left[\left(\frac{\partial V_{xc}}{\partial n}\right)_{\substack{n = n_0(\vec{r}) \\ \omega=0}}\right] n^{(2)}(\vec{r}, \omega) + \left[\frac{1}{2}\left(\frac{\partial^2 V_{xc}}{\partial n^2}\right)_{\substack{n = n_0(\vec{r}) \\ \omega=0}}\right] \left[n^{(1)}(\vec{r}, \omega)\right]^2 . \quad (3.103)$$

This formulation is completely equivalent to the modified Sternheimer method described in the previous section. Self-consistency in screening requires that the integral equation for δn, (3.91), be solved self-consistently, and is usually accomplished in 8-10 iterations. The main computation is reduced to that of finding the Green's function. The steps involved in the angular momentum decomposition of (3.97) are analogous to those described in the previous section.

The methods described in this and the previous sections are entirely equivalent, and comparable in the computational effort involved. However, there is one aspect in which the modified Sternheimer method has an advantage. This has to do with schemes for correcting for self-interaction in LDA, which is discussed in the next section.

3.5 Self-Interaction Corrections

As mentioned in Chapter 1, the use of the total electronic density in the calculation of the Coulomb, local-exchange, and local-correlation potentials results in the spurious inclusion of self-interaction terms. Among the undesirable consequences of this error are:

- The long range behavior of the one-electron potential for an ion of charge Q is $-Q/r$ instead of the correct $-(Q+1)/r$.

- The experimentally stable negative ions (e.g., H^-, F^-, etc.) are predicted to be unstable.

- Atoms and ions with one or two electrons (H, He, Li^+) are treated as having an exchange energy. In fact, no exchange energy is present in these systems.

- The calculated polarizabilities are systematically larger than observed values.

It is not our purpose here to discuss all the schemes that have been proposed to overcome these deficiencies. They have been reviewed, for instance, by Callaway and March (1984) and Perdew and Zunger (1981). Instead, we will only summarize here the main ideas leading up to one scheme that has been tested in the context of polarizability calculations.

First we need to introduce some notation for the electron density.

- $n(\vec{r})$ is the total density of all the electrons;
- $n_\sigma(\vec{r})$ is the density of electrons with spin component $\sigma = (\uparrow, \downarrow)$; $n(\vec{r}) = n_\uparrow(\vec{r}) + n_\downarrow(\vec{r})$.
- $n_{j\sigma}(\vec{r}) = |\psi_{j\sigma}(\vec{r})|^2$ is the particle density from a single-electron state of orbital j and spin σ; thus, $n_\sigma = \sum_j n_{j\sigma}$.

The self-interaction correction (SIC) affects two terms in the self-consistent potential of the electron. One term is the Hartree potential for the Coulomb energy. The Coulomb interaction is between pairs of electrons. A particle does not interact with itself. The proper Hartree potential for an electron in the state (j, σ) is

$$V_{H,j\sigma}(\vec{r}) = 2\int \frac{d^3r'[n(\vec{r}\,') - n_{j\sigma}(\vec{r}\,')]}{|\vec{r} - \vec{r}\,'|}. \tag{3.104}$$

All models of SIC use this form for the Hartree potential.

The exchange-correlation potential must also be altered in order to account for SIC. There have been different proposals for doing this. In order to keep things simple, the discussion is focused on just the exchange energy. However, similar ideas are easily extended to the correlation energy.

The exchange energy exists between electrons of identical spin components. The exchange energy of the homogeneous electron gas should not be written as being a function of the total density, but instead of the density of electrons of the same component of spin

$$\varepsilon_{x\sigma} = -\frac{3}{2\pi}(6\pi^2 n_\sigma(\vec{r}))^{1/3}. \tag{3.105}$$

This result should be compared with (2.21). They are identical in a system with no spin polarization, so that $2n_\sigma = n$. The two expressions are different for a system with spin polarization. The above formula is correct for a spin-polarized electron gas.

The exchange energy per electron is found for the homogeneous electron gas. There, each electron has its wave function spread uniformly throughout the system of volume V. Each electron contributes a term of order $(1/V)$ to the density n_σ. We can make a self-interaction correction by subtracting the density of the electron from the total density,

$$n_\sigma(\vec{r}) \rightarrow n_\sigma(\vec{r}) - |\psi_{\vec{k}\sigma}|^2 = n_\sigma(\vec{r}) + O(1/V), \tag{3.106}$$

$$\varepsilon_{x\sigma}(n_\sigma - |\psi_{\vec{k}\sigma}|^2) = \varepsilon_{x\sigma}(n_\sigma) + O(1/V). \tag{3.107}$$

The SIC has a negligible effect on the exchange and correlation energy of the homogeneous electron gas.

A very different answer is obtained for electron states which are localized on atoms or ions. Then, the charge density $n_\sigma - n_{j\sigma}$ is very different from n_σ. Typical examples are:

- s-states may have one electron of each spin component. Then $n_\sigma - n_{j\sigma} = 0$, and the density of other electrons in the same orbital with the same spin is zero.
- p-states have three electrons with the same spin in the same orbital. The SIC changes three to two electrons, which changes the answer by 33 %.

In the scheme where self-interaction correction for exchange is made via the use of the correct density, one uses

$$E_x = \int d^3r \sum_{j\sigma} n_{j\sigma}(\vec{r})\varepsilon_{x\sigma}(n_\sigma(\vec{r}) - n_{j\sigma}(\vec{r})). \tag{3.108}$$

The exchange potentials to use in the Kohn-Sham equations are found by a functional derivative,

$$V_{x,j\sigma}(\vec{r}) = \frac{\delta E_x}{\delta n_{j\sigma}} = \varepsilon_{x\sigma}(n_\sigma - n_{j\sigma}) - \frac{n_{j\sigma}(6\pi)^{1/3}}{2\pi} \sum_{\ell \neq j} [n_\sigma - n_{\ell\sigma}]^{-2/3}. \tag{3.109}$$

The exchange potential has many terms, which makes it cumbersome to use. It does give good results. One way to simplify this potential is to approximate it by the expression

$$V_{x,j\sigma} \approx -\frac{2}{\pi}(6\pi)^{1/3}[n_\sigma - n_{j\sigma}]^{1/3}. \tag{3.110}$$

This formula is easier to evaluate numerically. However, it is not derived as a functional derivative of the ground state energy. Mahan (1980) used this simplified expression to evaluate the ionization potential of some negative ions. These results will be discussed below. Another method was suggested by Lindgren (1971). His SIC for exchange was accomplished by subtracting the exchange energy for the one-electron state:

$$\varepsilon_{x,j\sigma} = \varepsilon_x(n) - \varepsilon_x(n_{j\sigma}), \tag{3.111}$$

$$V_{x,j\sigma}(\vec{r}) = -\frac{2}{\pi}(3\pi^2)^{1/3}[n(\vec{r})^{1/3} - n_{j\sigma}(\vec{r})^{1/3}]. \tag{3.112}$$

This procedure is different from evaluating the exchange energy for the reduced density.

The Lindgren procedure for the exchange energy was extended to the correlation energy by Perdew and Zunger (1981) as described below. In the version of the LDA generalized to a spin-polarized system (called the Local Spin Density approximation, or LSD), one approximates the exchange-correlation energy by that of a homogeneous, spin-polarized electron gas at the local values of the up and down electron densities. Equation (3.21) is generalized to read

$$E_{xc}[n_\uparrow, n_\downarrow] = \int d^3r\, n(\vec{r})\epsilon_{xc}(n_\uparrow(\vec{r}), n_\downarrow(\vec{r})), \tag{3.113}$$

where $\epsilon_{xc}(n_\uparrow, n_\downarrow)$ is the exchange-correlation energy per particle of the homogeneous electron gas at uniform densities $n_\uparrow, n_\downarrow$. Perdew and Zunger (1981) replace (3.113) by

$$E_{xc}{}^{SIC} = E_{xc}[n_\uparrow, n_\downarrow] - \sum_{j\sigma} E_{xc}[n_{j\sigma}, 0]. \tag{3.114}$$

This subtracts out the self-interaction part of the exchange-correlation energy of the electrons in each orbital. The minimization of the total energy functional now leads to the self-interaction corrected Kohn-Sham equations,

$$\{-\nabla^2 + V_{j\sigma}^{eff}\}\psi_{j\sigma} = \lambda_{j\sigma}\psi_{j\sigma}, \tag{3.115}$$

where the orbital-dependent effective potential is given by

$$V_{j\sigma}^{eff}(\vec{r}) = U(\vec{r}) + 2\int d^3r' \frac{n(\vec{r}\,') - n_{j\sigma}(\vec{r}\,')}{|\vec{r} - \vec{r}\,'|} + V_{xc}([n_\uparrow, n_\downarrow]; \vec{r}) - V_{xc}([n_{j\sigma}, 0]; \vec{r}). \tag{3.116}$$

Table 3.4: Ionization energies of negative ions in Rydbergs using self-interaction corrected LDA

ion	exp.	M[a]	PZ[b]
F^-	0.250	0.245	0.265
Cl^-	0.266	0.265	0.279
Br^-	0.247	0.266	

[a] G.D. Mahan, Phys. Rev. B **22**, 3102 (1980)
[b] J.P. Perdew and A. Zunger, Phys. Rev. B **23**, 5048 (1981)

The corrections to the direct Coulomb and exchange parts have been discussed above already. For the correlation potential also the self-interaction contribution due to the a single unpaired electron in the orbital in question is subtracted from the total.

Table 3.4 compares the ionization energy of negative ions calculated using the above two methods. They are obtained by calculating the total ground state energy for the state with N electrons, and with $N-1$ electrons, and then subtracting these two ground state energies. This number is very different than the eigenvalues. Mahan (1980) used the scheme of calculating the potential at a reduced density. Perdew and Zunger (1981) used the scheme of subtracting out the self-potential. Both methods of doing the SIC predict the negative ions are stable, and give accurate values for the ionization energy. However, this comes at the price of additional computational burden. The self-consistent potential is different for each orbital, and the orbitals corresponding to different eigenvalues are not orthogonal. Perdew and Zunger have shown that the latter does not introduce any serious error, and it is customary not to worry about the lack of orthogonality of the orbitals.

In Figure 3.5 we show the potentials (multiplied by r) for the $2p$ valence electrons in the Ne atom with and without self-interaction correction calculated with the Perdew-Zunger scheme. Note that the SIC potential correctly shows the $1/r$ behavior at large distances. In Table 3.5 we compare the $1s$, $2s$, and $2p$ eigenvalues for Ne with and without SIC computed using the Ceperley-Alder correlation energy. One sees that for the outer orbitals, the differences are substantial. The SIC calculations lead to a more tightly bound atom as a consequence of having the correct asymptotic potential.

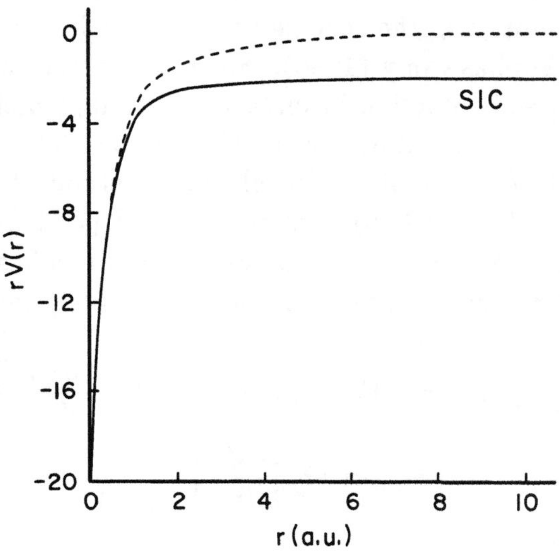

Figure 3.5: Effect of SIC on $2p$ orbital potential for Ne. The Perdew-Zunger scheme was used. Note that $r \times V(r)$ is shown.

Table 3.5: Orbital eigenvalues for Ne with and without SIC. The Perdew-Zunger SIC scheme was used.

Orbital	LDA	LSD-SIC
$1s$	60.51	65.34
$2s$	2.639	3.315
$2p$	0.995	1.684

We now consider how the above scheme affects the calculation of the polarizability. It is evident from the discussion of the previous paragraph that the polarizabilities calculated with SIC are apt to be smaller than those calculated in simple LDA. The more tightly bound valence electrons are less susceptible to external perturbations. However, there are certain conceptual difficulties in applying the scheme to the screening terms, as discussed by Senatore and Subbaswamy (1986). Since the potential is now orbital-dependent, the Green's function formalism is not applicable; it is not clear how to construct a Green's function in this context. Therefore, we discuss SIC in the context of the modified Sternheimer method of Section 3.3.

The potentials $V^{(\alpha)}$ of (3.46) become orbital dependent. The modification to the ground state potential has already been indicated above. The first-order self-consistent perturbation $V_{SCF,M}$ of (3.50) changes to

$$\begin{aligned} V^{j}_{SCF,M}(\vec{r}) &= V_{SCF,M}(\vec{r}) - 2\int d^3r' \frac{n_j(\vec{r}\,')}{\mid \vec{r}-\vec{r}\,' \mid} \\ &\quad - n_j(\vec{r}) \left[\frac{\partial V_{xc}}{\partial n}\right]_{n=n_j(\vec{r})} . \end{aligned} \tag{3.117}$$

The first term, given in (3.50), is the potential calculated with the total induced density; the second term corrects for the self-interaction contribution in the Hartree term; and the last term corrects for the same in the exchange-correlation term. The last term has cusps at the nodes of the orbital under consideration, and leads to an unphysical potential, as will be discussed in the next chapter.

As already mentioned, because of the orbital dependent potentials, the various ground state orbitals are no longer orthogonal. However, the work of Perdew and Zunger (1981) indicates that the degree of non-orthogonality is very small. Further, for isolated atoms the changes in orbital energies resulting from orthogonalization are practically negligible. Thus, in practice, one usually ignores the lack of orthogonality. As a practical matter, then, the numerical procedure discussed in Section 3.3 is applied here as well, with the only difference being that the potentials for the different orbitals are different. In the case of isolated atoms or ions, this leads to N self-consistent equations, where N is the number of occupied Kohn-Sham orbitals. An evaluation of SIC in the calculation of polarizabilities will be carried out in the next chapter.

Applying the SIC increases the complexity of the numerical calculation. One must calculate and store a separate potential $V_j(\vec{r})$ for each different localized state j. Generally one avoids this extra effort whenever possible. There are two situations where the SIC is worth the effort. One is for the calculations of negative ions. They are not bound in a LDA calculation. SIC must be used in order to obtain a bound ion. The other case where SIC provides more accuracy is in atoms or ions from the first row of the periodic table, where the electrons are in the 1s and 2s shells. There is no exchange within an s-shell, and better results are obtained when it is omitted. These two cases are important, so that it is useful to have a computer code with SIC in order to do calculations. However, for atoms whose outer shells are p- or d-electrons, SIC provides little improvement in the results for a great increase in calculational effort, and is generally avoided.

Chapter 4

Atoms and Molecules

4.1 Linear Polarizability

The dipole polarizability of atoms was the first quantity calculated using TDLDA. Almost simultaneous calculations were done by Stott and Zaremba (1980), Zangwill and Soven (1980), and Mahan (1980). The former two used the Green's function method, while Mahan used the modified Sternheimer Method. All three groups obtained similar results. Mahan also calculated the quadrupole and octupole polarizabilities. Zangwill and Soven calculated the optical absorption, which is related to the imaginary part of the frequency-dependent polarizability. The finite frequency results are discussed in the next section. Since that initial work, many more calculations have been reported of static and dynamic polarizability.

The method of calculating the polarizability was discussed in the prior chapter. The major steps in the calculation are summarized here for the Mth polarizability: $M = 1$ is dipole, $M = 2$ is quadrupole, $M = 3$ is octupole, etc. The calculation is equally easy for any value of M.

Recall from (2.96) that the ground state energy in the presence of a small, static external field $\mathcal{E}_M$ may be expanded as

$$E_N(\mathcal{E}_M) = E_N - \frac{1}{2!}\alpha_M \mathcal{E}_M^2 - \frac{1}{4!}\gamma_M \mathcal{E}_M^4 \ldots \tag{4.1}$$

The coefficients α_M and γ_M are the linear and nonlinear polarizability, respectively. This section is devoted to the calculation of the linear polarizability α_M. The polarizability is also defined as the constant of proportionality between the induced moment p_M and the applied field $\mathcal{E}_M$,

$$p_M = \alpha_M \mathcal{E}_M + O(\mathcal{E}_M^3). \tag{4.2}$$

The two definitions (4.1) and (4.2) are entirely equivalent for static fields. Equation (4.2) is used to calculate the polarizability, since we actually calculate the induced moment.

The static polarizability is a property of the ground state of the atom. According to (4.1) the ground state energy in the external field depends upon the polarizability. This important point provides the justification for the method of calculation. LDA is a method of calculating the ground state properties of electronic systems. LDA and TDLDA can be used to calculate the static polarizability accurately since it is a ground state property. The theoretical procedure is to calculate the change in the ground state energy when the field is applied. It is equivalent to the calculation of the induced moment (4.2). For the frequency-dependent case the justification is less rigorous, and involves an adiabatic approximation as discussed in Section 2.4. The applied field $\mathcal{E}_M$ is included in the Hamiltonian as a potential energy term of the form

$$V_M(\vec{r}) = e\mathcal{E}_M r^M P_M(\nu), \tag{4.3}$$

where $\nu = \cos(\theta)$ and e is the electron charge. A familiar example is the application of a static electric field. This case has $\mathcal{E}_M = -\mathcal{E}$ and $M = 1$ which gives $V_1 = -e\mathcal{E}r\nu$. The above formula is the generalization for higher values of M. The applied field is viewed as infinitesimal.

The applied field causes small distortions in the Kohn-Sham eigenfunctions. Only first-order changes are needed for the calculation of the linear polarizability. The first-order changes in the eigenfunctions and density are denoted with a superscript (1):

$$\begin{aligned} \psi_j(\vec{r}, \mathcal{E}_M) &= \psi_j(\vec{r}) + e\mathcal{E}_M \psi_j^{(1)}(\vec{r}) + \ldots, & (4.4) \\ n(\vec{r}, \mathcal{E}_M) &= n(\vec{r}) + e\mathcal{E}_M n_M^{(1)}(\vec{r}) + \ldots, & (4.5) \\ n_M^{(1)} &= 2\sum_j \Re\{\psi_j^* \psi_j^{(1)}\}. & (4.6) \end{aligned}$$

According to the theorem on angular momentum in Section 3.3, the change in density for an atom with only filled shells has its angular dependence determined by $n^{(1)}_M(\vec{r}) = n^{(1)}_M(r)P_M(\nu)$. The calculation of the linear polarizability is obtained easily for this case:

$$\begin{aligned}
p_M &= \int d^3r V_M(\vec{r}) n(\vec{r}, \mathcal{E}_M), \\
&= e^2 \mathcal{E}_M \int d^3r \; r^M P_M(\nu)^2 n^{(1)}_M(r), \qquad (4.7) \\
\alpha_M &= \frac{4\pi e^2}{2M+1} \int_0^\infty dr r^{M+2} n^{(1)}_M(r). \qquad (4.8)
\end{aligned}$$

In atomic units $e^2 = 2$. The polarizability is obtained by an integral over r after the calculation of the change in density $n^{(1)}_M(r)$. The change in density is obtained by solving for the change in the Kohn-Sham eigenfunctions. They obey the first-order differential equation

$$\begin{aligned}
[H_0 - \lambda_j]\psi^{(1)}_j(\vec{r}) &= -[v_{\text{SCF,M}}(r)P_M(\nu) - \lambda^{(1)}_j]\psi_j(\vec{r}), \qquad (4.9) \\
v_{\text{SCF,M}}(r) &= r^M + \frac{8\pi}{3} \int r'^2 dr' \frac{r_<}{r_>^2} n^{(1)}_M(r') + n^{(1)}_M(r) \frac{\partial V_{xc}}{\partial n_0}, \qquad (4.10) \\
\lambda^{(1)}_j &= \int d^3r v_{\text{SCF,M}}(r) P_M(\nu) |\psi_j(\vec{r})|^2. \qquad (4.11)
\end{aligned}$$

The equation for $v_{\text{SCF,M}}(r)$ is given previously in (3.55). Section 3.3 describes the method for solving the inhomogeneous differential equation in (4.9). Only the radial part of the function has to be obtained numerically. Its equation is:

$$\left[-\frac{\partial^2}{\partial r^2} + \frac{l'(l'+1)}{r^2} + V(r) - \lambda_{nl}\right](rR^{(1)}_{nl'}(r))$$

$$= -[v_{\text{SCF,M}}(r) - \lambda^{(1)}_j \delta_{ll'}] r R_{nl}(r), \qquad (4.12)$$

$$n^{(1)}_M(r) = \sum_{nl} \sum_{l'=|M-l|}^{l+M} D(l, l'; M) R^{(1)}_{nl'}(r) R_{nl}(r). \qquad (4.13)$$

where R_{nl} is the radial part of the Kohn-Sham eigenfunction while $R^{(1)}_{nl'}$ is the radial part of $\psi^{(1)}_j$. The quantum numbers (nl) denote the atomic shells, where l is the angular momentum. The coefficients $D(l, l'; M)$ are listed in Table 3.3. The numerical procedure is to solve (4.12) using the methods described in Section 3.3. Then one finds $n^{(1)}_M$ and the linear polarizability is found from the integral in (4.8).

It is worth pointing out the relationship between the standard perturbation theoretic expression,

$$\alpha = 2e^2 \sum_{\beta,j} \frac{|<j|z|\beta|>^2}{\lambda_\beta - \lambda_j}, \tag{4.14}$$

and the LDA form (4.8). In (4.14), which is merely the static limit of (2.77), the summation over electron states includes the occupied states j and the unoccupied states β. Dirac notation has been used for the matrix element

$$<j|z|\beta> = \int d^3r \psi_j^*(\vec{r}) z \psi_\beta(\vec{r}). \tag{4.15}$$

It is identical to the result obtained when we use as the perturbation $z = rP_1(\nu)$ instead of $v_{\mathrm{SCF},1}(r)\nu$. In order to show this identity, rewrite the summation by writing out one of the matrix elements. Then define $\psi_j^{(1)}$ to be the summation over the states ψ_β:

$$\sum_{\beta,j} \frac{|<j|z|\beta>|^2}{\lambda_\beta - \lambda_j} = \sum_j \int d^3r \psi_j^*(\vec{r}) z \psi_j^{(1)}(\vec{r}),$$

where

$$\psi_j^{(1)}(\vec{r}) = \sum_\beta \frac{\psi_\beta(\vec{r}) <\beta|z|j>}{\lambda_\beta - \lambda_j}.$$

The first-order change in the wave function $\psi_j^{(1)}$ is the same one defined in (2.76). The first-order change in the density is

$$n_1^{(1)} = 2\Re\{\sum_j \psi_j^*(\vec{r}) \psi_j^{(1)}(\vec{r})\}. \tag{4.16}$$

Thus, (4.14) reduces to

$$\alpha = e^2 \int d^3r z n_1^{(1)}(\vec{r}). \tag{4.17}$$

This is merely (4.9), which is the one calculated in the Sternheimer formalism. This expression is the LDA version of the polarizability, since the field is not calculated self-consistently.

The standard expression for the polarizability is expressed as a summation over the unoccupied states ψ_β. It may appear as if the polarizability

Table 4.1: Dipole polarizabilities in atomic units (a_0^3). Experimental results in the first column. The last four columns are the results of a TDLDA calculation using four kinds of correlation: Gunnarsson-Lundqvist, Vashishta-Singwi, Gordon-Kim, and Ceperley-Alder.

Ion	Exp	GL	VS	GK	CA
Ne	2.663[a]	2.99	3.07	3.07	3.15
Na^+	0.998[b]	1.07	1.08	1.08	1.09
Mg^{++}	0.493[c]	0.51	0.51	0.52	0.52
Ar	11.08[a]	11.80	12.08	12.03	12.48
K^+	5.47[b]	5.70	5.77	5.76	5.86
Ca^{++}	3.26[c]	3.34	3.36	3.36	3.40
Kr	16.74[a]	17.70	18.14	18.04	18.86

[a] R.R. Teachout and R.T. Pack, Atomic Data **3**, 196 (1971)
[b] R.R. Freeman and D. Kleppner, Phys. Rev. A **14**, 1614 (1976); J.C. Lombardi, Phys. Rev. A **36**, 1445 (1987)
[c] H. Eissa and U. Öpik, Proc. Phys. Soc. (London) **92**, 556 (1967)

is not a property of the ground state, since the definition involves unoccupied electron states. However, it was shown earlier in this section that the static polarizability is indeed a property of the ground state. The states ψ_β are actually properties of the ground state, since they are eigenstates of the Hamiltonian H_0 of the ground state. The potential energy term in H_0 is calculated with the electron density of the ground state. The polarizability is indeed a property of the ground state of the atom.

Excited states of atoms can be polarized, and they have a polarizability. These quantities can not be found rigorously using LDA, since they are not properties of the ground state.

4.1.1 Atoms and Positive Ions

The present method is most accurate for atoms or ions with all electrons in filled atomic shells. Table 4.1 shows numerical results for dipole polarizabilities calculated with the present method. The first column shows experimental values where reliable data are available. The next four columns are the calculated results using the four theories for correlation reviewed in Section 3.1: Gunnarsson-Lundqvist (GL), Vashishta-Singwi (VS), Gordon-Kim

(GK), and Ceperley-Alder (CA). Identical results for GL correlation were found by Stott and Zaremba (1980) using Green's function methods. The LDA results are systematically larger than experimental values. This is due to the poor treatment of the long-distance behavior of the potential in LDA. The effect of self-interaction correction will be discussed in the next section.

The dipole polarizability has the dimensional units of volume. For atoms the polarizability scales, approximately, with the volume of the atom. Big atoms have big polarizabilities while small atoms have small ones. The GL correlation energy is the largest in magnitude, and has the strongest correlation potential. Electrons are more bound with this potential. The atom is smaller in size. The polarizability is the smallest of the four values. The other choices for correlation produce electron states with less binding energy. Their eigenfunctions extend farther, and their polarizabilities are larger. Our estimate is that the atomic polarizability is about 7% too high when calculated using GL correlation. This error of 7% arises consistently from atoms with few electrons such as neon to those with many such as krypton. The error is even higher when using the other three choices for correlation. The best polarizabilities are obtained with GL correlation. Their correlation energy may not be the best one for the homogeneous electron gas. One can conclude, however, that it gives the better correlation potential for atoms.

Here the computational reality departs from the initial philosophy. The original idea was that the best exchange-correlation potential is found by taking the exact result for the homogeneous electron gas and replacing the Fermi wave vector k_F by the local function of density $(3\pi n(\vec{r}))^{1/3}$. However, that procedure results in a potential energy for electrons in atoms which is too small in magnitude. The electrons are not sufficiently bound. The polarizability is a sensitive measure of atom size. The Gunnarsson-Lundqvist correlation potential is better because it produces atoms with the most binding compared to the other correlation potentials. Even better results would be obtained with a stronger correlation potential. The empirical preference for obtaining good results outweighs the esthetic virtue of using the best correlation potential for the homogeneous electron gas. It might be pointed out here that this was the motivation for the adjustable α in Slater's Xα method.

Table 4.2 shows the values of dipole polarizability calculated by the Hartree-Fock approximation, and by the coupled Hartree-Fock approximation. The latter method is equivalent to making the applied potential self-consistent. The first column shows experimental results. The second column shows results of Langhoff and Hurst (1965) using the HFA. These authors

Table 4.2: Dipole polarizabilities (a_0^3) calculated using the Hartree-Fock approximation. The first column is experimental values. The second column is results using HFA. The last four columns are results using TDHFA, which is also called the coupled-Hartree-Fock approximation.

Ion	exp.	(*d*)	(*e*)	(*f*)	(*g*)	(*h*)
Ne	2.663[a]	2.82	2.36	2.35	2.377	
Na^+	0.998[b]	1.11	0.945	0.939	0.945	0.965
Mg^{++}	0.493[c]	0.55	0.47		0.470	0.476
Ar	11.08[a]	16.61	10.0	9.65	10.76	
K^+	5.47[c]	7.69	5.32	4.91	5.461	5.216
Ca^{++}	3.26[c]	4.39	3.21		3.261	3.057

[a] R.R. Teachout and R.T. Pack, Atomic Data **3**, 195 (1971)
[b] R.R. Freeman and D. Kleppner, Phys. Rev. A **14**, 1614 (1976)
[c] H. Eissa and U. Öpik, Proc. Phys. Soc. (London) **92**, 556 (1967)
[d] P.W. Langhoff and R.P. Hurst, Phys. Rev. **139**, A1415 (1965)
[e] J. Lahiri and A. Mukherji, Phys. Rev. **153**, 386 (1967); **155**, 24 (1967)
[f] C. Litt, Phys. Rev. A **7**, 911 (1973)
[g] R.P. McEachran, A.D. Stauffer, and S. Greita, J. Phys. B **12**, 3119 (1979)
[h] P.C. Schmidt, A. Weiss, and T.P. Das, Phys. Rev. B **19**, 5525 (1979)

also have a fine compilation of previous calculations. The last four columns show results using TDHFA. There is a significant reduction in the values between HFA and TDHFA, just as there is a reduction between LDA and TDLDA. The TDHFA results are all too small compared with experiment. The values vary a bit, but seem to be about 5% too low. Since TDLDA is about 7% too high, the best estimate is obtained by averaging the results from the two theoretical methods.

How are the experimental polarizabilities determined? The answer is different for each series of atoms. All results are from atomic physics. Teachout and Pack (1971) have compiled the experimental results for neutral atoms as obtained by various techniques. For the rare gas atoms the best method seems to be the variation of the refractive index with particle density. The Clausius-Mossotti relation

$$\alpha = \frac{3V_0}{4\pi}\frac{n^2-1}{n^2+2} \tag{4.18}$$

gives the dipole polarizability as a function of the volume per atom V_0 and the refractive index n.

Values for the alkali ions come from studying energy levels of highly excited Rydberg states of the outer electron. Consider the system of an Na^+ ion with closed electronic shells plus an electron. For Rydberg states with nonzero values of angular momentum l, the centrifugal potential pushes the outer electron away from the ion core. The electrostatic potential from the Na^+ ion which acts upon the outer electron can be approximated as

$$V(r) = -\frac{e^2}{r} - \frac{e^2\alpha_d}{2r^4} - \frac{e^2\alpha_q}{2r^6} - \ldots \tag{4.19}$$

where α_d, α_q are the dipole and quadrupole polarizabilities. The first term is from the Coulomb potential, while the other terms arise from the polarizability of the ion. Without the polarizability terms the eigenvalues would be the Rydberg series of the hydrogen atom for large values of l. The polarizability terms cause small deviations from the Rydberg series. These deviations are measured carefully, and used to deduce an accurate value of the dipole polarizability α_d. The quadrupole polarizability is given less accurately.

Table 4.3 shows theoretical results for static dipole, quadrupole, and octupole polarizability calculated using TDLDA and also coupled-Hartree-Fock. For TDLDA the calculations are equally easy for any value of M. Mahan (1980) was the first to calculate the quadrupole and octupole polarizabilities this way. Stott, Zaremba, and Zobin (1982) used the Green's function method to find quadrupole results which are identical to those in Table 4.3. As already pointed out, screening is less important for $M \geq 2$ since the r^M factor is quite small in the interior of the atom. For TDHFA we present the results of Schmidt, Weiss, and Das (1979) who were the first to calculate atoms with high atomic number. Bartolotti (1984) suggested another LDA method based upon a variational calculation. It gives less accurate results than the present method.

The results for TDHFA are systematically smaller than those of TDLDA. For most ions the difference is about 10-20%. However, the difference is much larger for Cu^+ and Ag^+, where the quadrupole polarizabilities from the two methods differ by a factor of 2. No experimental results are available for

Table 4.3: Theoretical dipole, quadrupole and octupole polarizabilities in atomic units. The first three columns of numbers give results using TDLDA. The last two columns give results in the coupled-Hartree-Fock method. The latter values are systematically lower.

	TDLDA			TDHFA[a]	
Ion	α_d	α_q	α_o	α_d	α_q
Ne	2.99	9.35	61.7	2.60	6.42
Na^+	1.07	1.88	6.27	0.97	1.45
Mg^{++}	0.51	0.60	1.27	0.48	0.50
Ar	11.8	59.6	657	11.0	48.2
K^+	5.70	18.0	120.	5.22	16.12
Ca^{++}	3.34	7.48	35.1	3.06	7.01
Cu^+	7.68	59.2	281.	5.15	20.00
Zn^{++}	2.85	9.28	34.2	2.24	4.72
Kr	17.7	108.5	1445.	16.6	78.8
Rb^+	9.41	38.7	334.	8.67	36.1
Sr^{++}	5.95	18.5	117.	5.51	18.0
Ag^+	9.93	59.7	396.	8.27	27.7
Cd^{++}	5.30	19.6	94.9	4.70	12.5

[a] M.B. Doren, J. Phys. B **5**, L151 (1972); P.C. Schmidt, A. Weiss, and T.P. Das, Phys. Rev B **19**, 5525 (1979)

the quadrupole and octupole polarizability, except for a few values such as Na^+. A recent fit of Na^+ by Lombardi (1987) gave values of $\alpha_q = 1.66$ or 1.36 a.u. depending upon the fitting procedure. The TDHFA result is about 1.52 a.u. while the TDLDA result is 1.88 a.u. The best theoretical value is probably near the average of the two theories, which is close to the value of 1.66 found by one of Lombardi's fitting methods.

It is interesting to investigate whether there are scaling relations between the multipole polarizabilities for the same atom or ion. If one knows the dipole polarizability, can one predict the polarizabilities with higher values of M? After plotting on various kinds of graph paper, it is evident that the scaling is a power law. Figure 4.1 shows a plot of quadrupole polarizability *versus* dipole polarizability for different atoms and ions. The lines connect the atoms or ions with the same valence: $Z = 0$ for Ne, Ar, Kr; $Z = 1$ for Na^+, K^+, Rb^+, etc. The higher polarizabilities are given well for atoms with outer p-shells by the formulas

$$\alpha_q = \frac{2.07(\alpha_d)^{1.38}}{\sqrt{1 + 0.47Z}} \tag{4.20}$$

$$\alpha_o = \frac{8.9(\alpha_d)^{1.8}}{\sqrt{1 + 1.67Z}} \tag{4.21}$$

The various constants in these formulas are entirely empirical, and are obtained by a least-squares fit to the theoretical values.

The origin of the exponents 1.38 and 1.8 is mysterious. For example, a simple scaling idea is based upon length (L). Since the dipole polarizability has the units of L^3, and the quadrupole polarizability has the units of L^5, a scaling based on length would predict that $\alpha_q \sim \alpha_d^{5/3}$. However, the exponent of $5/3 = 1.67$ is far from the empirical value of 1.38, and scaling by length gives the wrong exponent.

The polarizabilities for ions with the outer electrons in filled d-shells are not given by these scaling formulas. These ions have polarizabilities which are slightly larger than predicted by the scaling formulas for electrons in p-shells. A separate scaling formula was not constructed for d-shells, since fews data points were available.

The local density approximation provides an easy method of calculating the polarizabilities of atoms or ions with all electrons in closed shells. When the applied field is found self-consistently, the dipole polarizabilities are about 7% too high when compared to experimental values. The best results are found using the correlation energy of Gunnarsson and Lundqvist.

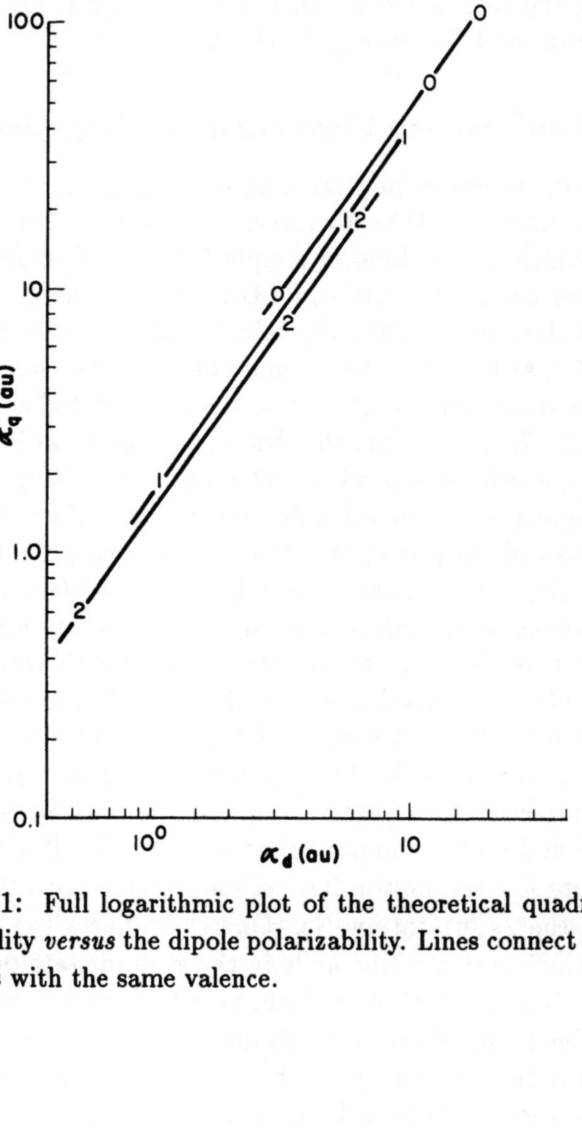

Figure 4.1: Full logarithmic plot of the theoretical quadrupole polarizability *versus* the dipole polarizability. Lines connect atoms or ion sets with the same valence.

The systematic numerical errors for quadrupole and higher polarizabilities are not known.

Probably the error increases with higher values of M. Values calculated in the coupled-Hartree-Fock approximation are systematically lower than the experimental values. Probably the best theoretical estimate is obtained by averaging the results from TDLDA and TDHFA. One is too high, the other is too low, and the average is about right!

4.1.2 Self-Interaction Correction and Negative Ions

So far we have discussed neutral atoms and positive ions, which are predicted to be stable in LDA. However, many closed-shell negative ions are also stable, which do not bind in simple LDA. As discussed in Section 3.5, self-interaction correction (SIC) in LDA leads to the correct treatment of negative ions. Hence, polarizability calculations for free negative ions require the use of SIC, at least for the ground state. To be completely consistent, SIC should be incorporated into the screening potential as well, as discussed in Section 3.5. In particular, the first-order potential is given in (3.117), and is explicitly orbital-dependent. One problem arising out of this fact is the nonorthogonality of the orbitals belonging to different eigenvalues. We have mentioned already that this is not a serious problem; the degree of nonorthogonality is small, and is usually neglected. One may always adopt an orthogonalization procedure, if desired. However, there is a more serious problem with extending SIC to the perturbed orbitals which has to do with artificial singularities introduced into the screening potential. We discuss this point before turning to negative ion polarizabilities.

It was mentioned in Section 3.5 that the exchange part of the self-interaction correction for the screening potential has singularities at the positions of the nodes of the unperturbed orbitals. To illustrate this point we show in Figure 4.2 the first-order, self-interaction corrected self-consistent potential for the $2s$ orbital for Ne. One clearly sees the singularity in the potential at the location of the node in the ground state orbital. As already discussed this is an artifact of the local density form for the exchange energy adopted in the LDA. There is no simple way to get rid of it. One option is to note that the singularity in the potential is fairly mild, having very little effect on the first-order orbital, and calculate the polarizability in spite of the known problem with the potential. As an example, the Ne $2s$ first-order orbital calculated with the singular screening potential (solid line) is compared in Figure 4.3 with that calculated without SIC in the screening

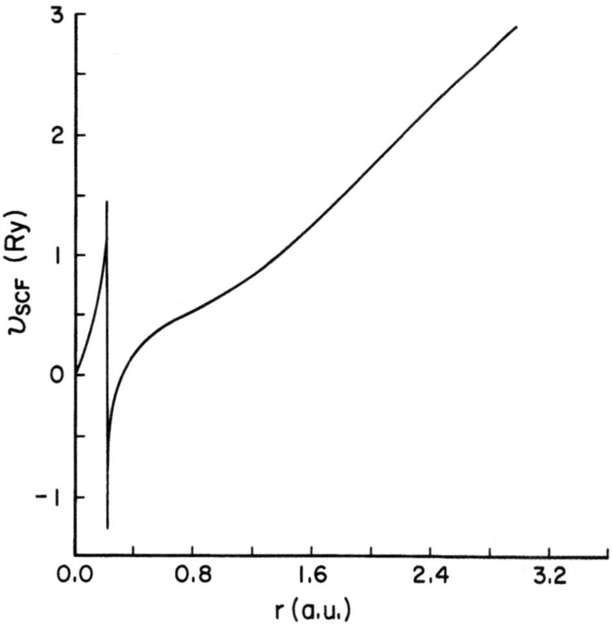

Figure 4.2: The self-consistent perturbation potential for the 2s orbital of Ne, with SIC. The singularity occurs at the node of the unperturbed orbital.

potential (dashed line). Another option is to just turn off the problematic exchange contribution to the self-interaction correction of the screening potential. Yet another option, which we adopt in our discussion of free negative ions, is not to perform any self-interaction correction in the screening term.

To assess the magnitude of the effect of SIC in the calculation of the polarizability we present in Table 4.4 the results for rare gas atoms using some of the options mentioned above. The values are taken from Senatore and Subbaswamy (1986), and were calculated with Ceperley-Alder correlation, and Perdew-Zunger self-interaction correction scheme. Similar results were also obtained by Böbel, Longinotti, and Fumi (1987), although they did not consider the effect of the singularity in the potential. FSIC refers to full SIC, namely for unperturbed and perturbed quantities. PSIC refers to partial SIC, namely for unperturbed quantities only. Note that as expected, use of SIC in the unperturbed state leads to more tightly bound atoms, and hence to lower polarizabilites than in LDA as can be seen by comparing the columns labeled LDA and PSIC. This effect is masked somewhat by the

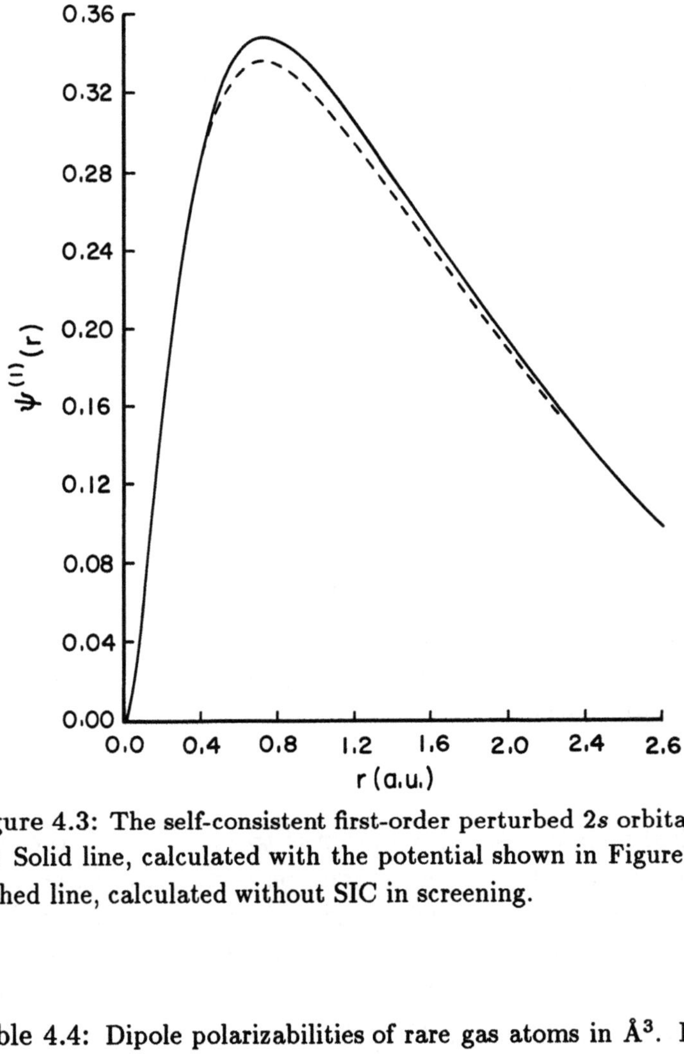

Figure 4.3: The self-consistent first-order perturbed 2s orbital for Ne: Solid line, calculated with the potential shown in Figure 4.2; dashed line, calculated without SIC in screening.

Table 4.4: Dipole polarizabilities of rare gas atoms in Å^3. LDA is without self-interaction correction (SIC) at any stage. FSIC is with full SIC, and PSIC is partial SIC, as described in the text.

Atom	Expt.	LDA	FSIC	PSIC
He	0.205	0.246	0.191	0.195
Ne	0.396	0.452	0.367	0.379
Ar	1.64	1.77	1.67	1.73
Kr	2.48	2.67	2.55	2.66
Xe	4.04	4.26	4.16	4.34

Table 4.5: Dipole polarizabilities of free, negative ions in Å^3. PSIC is partial SIC-LDA result; CHF is coupled Hartree-Fock; CHF-MP is CHF with many-body perturbation corrections.

Ion	PSIC	CHF	CHF+MP
H^-	12.0	13.3[a]	16.7[a]
F^-	1.80	1.58[b]	2.48[b]
Cl^-	5.47	4.66[b]	5.49[b]

[a]Fowler and Madden (1984)
[b]Fowler and Madden (1985)

orthogonalization problem in atoms other than He. However, now the calculated values tend to be lower than the experimental values. The present SIC scheme makes the atoms too rigidly bound. Comparison of the columns labeled FSIC and PSIC shows that the effect of doing a full SIC calculation in the perturbed quantities does not lead to as significant a difference. Of course, the FSIC calculation entails the singular potentials discussed above. The reason one gets a reasonable result in spite of the singularity in the potential is that the first-order orbital is only slightly modified by the weak singularity as shown for the $2s$ orbital of Ne in Figure 4.3. Nevertheless, we reject the use of SIC in the the screening term due to the unphysical nature of the potential. As already mentioned, the use of SIC is best avoided altogether except where absolutely necessary — as in the case of free negative ions. In such cases SIC should be used only for the unperturbed quantities where no unphysical singularities are involved.

We now turn to results of calculation for the stable, free negative ions with closed-shell configurations. SIC is used only in the unperturbed calculation. The results, labeled PSIC, are compared with Hartree-Fock results in Table 4.5. The results of coupled Hartree-Fock (CHF) with and without many-body perturbation corrections (MP) calculated by Fowler and Madden (1985) are also included for comparison. It is seen that the PSIC results are somewhat better than CHF; for Cl^- the PSIC result almost coincides with the CHF-MP result.

For the H^- ion the upper and lower bounds on the dipole polarizability are known: $\alpha = 30.5 \pm 0.4$ Å^3 (Glover and Weinhold, 1976). It is seen that the PSIC, as well as approximations based upon Hartree-Fock, seriously underestimate the polarizability of the H^- ion. Neutral hydrogen and the H^- ion are the worst cases for density functional schemes. The present SIC scheme gives an electron affinity for H which is twice the measured value.

We thus see that the LDA gives polarizabilities for atoms and ions that are in reasonable agreement with experiment, when self-consistent screening is taken into account. Simple LDA overestimates the polarizability, while our ground state self-interaction correction scheme makes the atoms and ions too compact. It should be noted, however, that these LDA schemes give results comparable to TDHFA (or CHF) results, with only a fraction of the effort involved.

4.1.3 Frequency Dependence

As shown in Chapter 3, the calculation of the frequency dependence of the polarizability in either the modified Sternheimer approach or the Green's function approach is a reasonably straightforward procedure. However, it should be clear that the results should only be reliable far away from resonances. Simple LDA, in particular, gives wrong eigenvalues for the value of the ionization threshold because of the incorrect long-range behavior of the potential. Thus, near a resonance, where the real part of the polarizability goes to infinity, the agreement with experimental values will be poor. The dispersion far away from resonance should be reproduced reasonably well by the theory, especially when self-interaction correction is applied for the unperturbed state. Frequency dependence has been tested in the case of rare gas atoms by Senatore and Subbaswamy (1986), and their results are shown in Table 4.6.

4.1.4 Molecules

Going beyond closed-shell atoms and ions invariably requires either additional approximations or a significant increase in computational effort. For open-shell atoms with partially filled p- or d-levels, for instance, the density is not spherically symmetric, and one usually proceeds with spherical averaging over magnetic quantum numbers. Besides, such atoms occur in gas phase only in molecular form, and there is no direct interest in calculating their atomic polarizabilities.

Levine and Soven (1984) performed the computation of polarizabilities of small molecules using time-dependent $X\alpha$ theory. They used the Green's

Table 4.6: Frequency dependence of the linear polarizability of rare gas atoms expressed as $\alpha(\omega) = \alpha_0(1 + C_2\omega^2)$. ω is in wave number units. α_0 are given in Table 4.4.

	C_2 (10^{-10} cm^2)	
	TDLDA	Expt.[a]
He	0.31	0.24
Ne	0.31	0.23
Ar	0.65	0.54
Kr	0.85	0.75
Xe	1.14	1.05

[a] P.J. Leonard (1974)

function formulation with a representation of the orbitals as well as the Green's function in a single-center expansion about the molecular center. The calculation of the induced perturbation potential is facilitated in this representation. However, the solution of the *ground state* problem in the single-center representation is very cumbersome. Hence, they performed the ground state calculation in a Gaussian basis representation, subsequently projecting needed quantities in a single-center representation. Construction of the Green's function requires the solution of coupled radial differential equations. They calculated the dipole polarizability tensor for the N_2 and C_2H_2 molecules. For these two molecules of C_{2v} symmetry there are two independent components of the dipole tensor: $\alpha_{||}$, for electric field parallel to the molecular axis, and $\alpha_{\perp}$ for field perpendicular to the molecular axis. Their results for photon energy of 2.5 eV are reproduced in Table 4.7. The LDA values result from the independent particle approximation, which neglects screening. The TDLDA results are very reasonable, and represent a great improvement over the LDA. Levine and Soven (1984) also treated the photoabsorption cross section by this method.

4.2 Photoionization

As already discussed in Chapter 2, the time-dependent local density approximation (TDLDA) can be used to calculate the response of an atom or solid to a time-dependent electric field. Here we consider the response of sin-

Table 4.7: Optical frequency polarizabilities of N_2 and C_2H_2 at 2.71 eV. Both independent components of the tensor are given. From Levine and Soven (1984).

	$\alpha_{\parallel}$ (Å^3)	$\alpha_{\perp}$ (Å^3)
N_2		
LDA[a]	5.10	2.35
TDLDA[a]	2.19	1.55
Expt.[b]	2.27	1.55
C_2H_2		
LDA[a]	27.30	4.24
TDLDA[a]	5.46	3.04
Expt.[b]	4.86	2.94

[a]Z.H. Levine and P. Soven, Phys. Rev. B **29**, 625 (1984)
[b]G.R. Alms, A.K. Burnham, and W. Flygare, J. Chem. Phys. **63** 3321 (1975)

gle atoms to electromagnetic waves. One quantity of interest is the dipole polarizability as a function of frequency $\alpha_d(\omega)$ which we have discussed previously, and compared with experiments (see Table 4.6). The polarizability is real for small frequencies. For large frequencies, the atom can absorb the radiation and photoionize an electron. Then the polarizability is complex. The imaginary part gives the photoionization cross section, which is given by

$$\sigma(\omega) = \frac{4\pi\omega}{c}\Im\{\alpha_d(\omega)\}. \tag{4.22}$$

The cross section has the dimensions of area, and is given in units of megabarns (Mb) which equal 10^{-18} cm^2. To derive (4.22) consider the LDA expression for the dipole polarizability as a function of frequency (see (2.77), (4.14)):

$$\alpha_d(\omega) = e^2 \sum_{im} |< i|z|m >|^2 f_i(1-f_m)\left\{\frac{1}{\lambda_{mi}+\omega+i\delta} + \frac{1}{\lambda_{mi}-\omega-i\delta}\right\}. \tag{4.23}$$

The quantity δ is infinitesimal. Equation (4.22) follows by comparing the imaginary part of the above equation with the golden rule expression for the absorption cross section,

$$\sigma(\omega) = \frac{4\pi^2 e^2}{c} \sum_{im} | < i|z|m > |^2 f_i(1 - f_m)\delta(\hbar\omega - \lambda_m + \lambda_i). \qquad (4.24)$$

The photoionization has a frequency threshold ω_T equal to the ionization energy (E_i) of the atom ($\hbar\omega_T = E_i$). The ionization energy is calculated very accurately for atoms using LDA. It is given by the difference of the ground state energy of the atom in the two charge states. One does an LDA self-consistent calculation for the atom with N electrons, and another calculation for the atom with $N - 1$ electrons. The threshold energy is the difference between these two ground state energies, $E_G(N - 1) - E_G(N) = E_i$. Notice that this quantity is *not* simply related to Kohn-Sham eigenvalues. Almbladh and von Barth (1985) have shown that the eigenvalue of the uppermost occupied density functional orbital equals the exact ionization potential. Simple LDA loses this feature; the self-interaction correction, which leads to the asymptotically correct potential, improves the accuracy.

The technique for calculating the photoionization spectra was developed by Zangwill and Soven (1980). It is successful, and has been widely adopted. This method is reported here. One interesting feature is that even though it gives the wrong threshold energy, given by the magnitude of the eigenvalue of the electrons in the outer shell of the atom, the technique gives accurate predictions for the cross sections in the range of experimental frequencies.

The photoionization has sharp absorption lines at frequencies less than threshold. They arise from transitions to excited states of the atom, in which the electron is still bound. As pointed out in Section 3.5, simple LDA has the wrong asymptotic potential, and does not predict the highly excited (unoccupied) bound states. Self-interaction correction leads to a great improved in the predicted threshold, while also picking up the bound to bound transitions.

4.2.1 Outgoing Waves

The dipole polarizability is calculated as a function of frequency using TDLDA. It is a complex quantity for frequencies larger than the ionization threshold. TDLDA is not a rigorous formalism for the calculation of properties at finite frequency. LDA is rigorous for ground state properties, but not for excited state properties. This issue has been discussed previously in Section 2.4.

Our discussion begins by proving a theorem. The results of this theorem have been utilized throughout the book. However, the need for the theorem is most evident when calculating phenomena at finite frequency. Consider the LDA expression for the dipole polarizability as a function of frequency in (4.23). This expression is not what we shall evaluate below since it has the perturbation z rather than V_{SCF}. However, this expression is useful for proving the theorem. The factor of f_i denotes that the initial state $|i>$ is occupied. The factor of $(1-f_m)$ denotes that the final state $|m>$ must be empty. The theorem is that the factor of $(1-f_m)$ may be ignored. This is not an approximation, but is an exact result. The importance of the theorem is shown by manipulating (4.23) using the same steps in equations (4.14) to (4.17). The integral in one matrix element is written out explicitly,

$$\alpha_d(\omega) = e^2\sum_i f_i \int d^3r\psi_i^*(\vec{r})z[\psi_i^{(1,+)}(\vec{r})+\psi_i^{(1,-)}(\vec{r})], \quad (4.25)$$

$$\psi_i^{(1,\pm)}(\vec{r}) = \sum_m (1-f_m)\frac{\psi_m(\vec{r})<m|z|i>}{\lambda_{mi}\pm(\omega+i\delta)}. \quad (4.26)$$

The first-order change in the wave function $\psi_i^{(1,\pm)}$ is evaluated by operating on both sides of the equation by $[H_0-\lambda_i\pm\omega]$, which gives

$$\begin{aligned}[H_0-\lambda_i\pm\omega]\psi_i^{(1,\pm)} &= \sum_m (1-f_m)\psi_m <m|z|i>, \\ &= z\psi_i(\vec{r}) - \sum_m f_m\psi_m(\vec{r})<m|z|i>. \quad (4.27)\end{aligned}$$

The last term on the right shows that one should subtract out the occupied orbitals. We have not been doing that in the modified Sternheimer method. The theorem we shall prove is that this term can be ignored without making an error. This result is fortunate, since including this term increases the numerical effort.

The theorem is that there is no need to restrict the summation over m in (4.27) to those states which are unoccupied. The factor of $(1-f_m)$ in (4.23) can be replaced by unity. The proof starts by taking the right-hand side of (4.23) and averaging it with the expression obtained by interchanging the summation variables i and m:

$$\begin{aligned}\alpha_d(\omega) = \frac{e^2}{2}\sum_{im} |<i|z|m>|^2 \Big\{ & f_i(1-f_m)\left[\frac{1}{\lambda_{mi}+\omega+i\delta}+\frac{1}{\lambda_{mi}-\omega-i\delta}\right] \\ & +f_m(1-f_i)\left[\frac{1}{\lambda_{im}+\omega+i\delta}+\frac{1}{\lambda_{im}-\omega-i\delta}\right]\Big\}. \quad (4.28)\end{aligned}$$

The energy denominators in the second square bracket equal the negative of those in the first square bracket. The expression may be simplified to

$$\begin{aligned}\alpha_d(\omega) &= \frac{e^2}{2}\sum_{im} | < i|z|m > |^2[f_i(1-f_m) - f_m(1-f_i)] \\ &\times \left[\frac{1}{\lambda_{mi}+\omega+i\delta} + \frac{1}{\lambda_{mi}-\omega-i\delta}\right], \qquad (4.29) \\ &= \frac{e^2}{2}\sum_{mi} | < i|z|m > |^2[f_i - f_m]\left[\frac{1}{\lambda_{mi}+\omega+i\delta} + \frac{1}{\lambda_{mi}-\omega-i\delta}\right].\end{aligned}$$

The terms with $f_i f_m$ cancelled. The term with f_m can be shown to equal the term with f_i by again interchanging dummy variables i and m. Thus we find the final expression

$$\alpha_d(\omega) = e^2 \sum_{im} f_i | < i|z|m > |^2 \left[\frac{1}{\lambda_{mi}+\omega+i\delta} + \frac{1}{\lambda_{mi}-\omega-i\delta}\right]. \qquad (4.30)$$

Comparing this expression with (4.23), the only difference is that the factor of $(1 - f_m)$ is missing. There is no longer a restriction that the final state m must be unoccupied. This feature seems counter-intuitive, but is correct. For example, in the argon atom the $3s$ and $3p$ shells are entirely filled. When calculating the polarizability, we will now include virtual transitions from $3s \rightarrow 3p$ which are positive and make the polarizability larger. However, we also include the transitions $3p \rightarrow 3s$ which are negative and make the polarizability smaller. They cancel exactly, and do not affect the final answer. This cancellation only occurs for the real part of the polarizability. For the imaginary part of the polarizability, we exclude unwanted transitions in the photoionization spectra by avoiding the discrete values of frequency where they occur. The restriction, that the final state be empty, can be omitted in the calculation of the real part of the polarizability. This makes the numerical techniques simpler. This proof also applies to the calculation of the static polarizability, where we used the results of the theorem in the calculations of the previous sections.

In TDLDA the inhomogeneous differential equation (4.27) is replaced by

$$\begin{aligned}[H_0 - \lambda_i \pm \omega]\psi_i^{(1,\pm)}(\vec{r}) &= V_{SCF}(\vec{r}, \pm\omega)\psi_i(\vec{r}), \qquad (4.31) \\ n^{(1)}(\vec{r}, \pm\omega) &= \sum_i f_i[\psi_i^*(\vec{r})\psi_i^{(1,\pm)}(\vec{r}) + \psi_i(\vec{r})\psi_i^{(1,\mp)*}(\vec{r})], \qquad (4.32) \\ \alpha_d(\omega) &= e^2 \int d^3r\, z n^{(1)}(\vec{r}, \omega). \qquad (4.33)\end{aligned}$$

The self-consistent potential was discusssed in Section 2.4. These equations are solved to obtain the photoionization spectra.

In solving (4.31) there are two possibilities. One is that the quantity $\Lambda_i = \lambda_i \pm \omega < 0$. Then the function $\psi_i^{(1,\pm)}$ decays to zero at large values of r. The equation is solved using exactly the same techniques discussed in Section 3.3. The only change is that the Λ_i is used in place of λ_i. The other possibility is that $\Lambda_i > 0$, so that photoionization can occur for electrons in the shell i. Then a new technique is required to solve this equation.

In photoionization, a photon from the ac electric field is absorbed by an electron in the atom. The electron is excited to a state of positive energy, and leaves the vicinity of the atom. Its wave function is an outgoing wave. In the present formalism, we need to solve (4.31) with the boundary condition appropriate for outgoing waves. How is that done?

The Green's function for the atomic orbitals can be used to write a formal solution to (4.31):

$$G(\vec{r}, \vec{r}', E) = \sum_j \frac{\psi_j(\vec{r})\psi_j^*(\vec{r}')}{\lambda_j - E - i\delta}, \tag{4.34}$$

$$\psi_i^{(1,-)}(\vec{r}) = \epsilon \phi_{\vec{k}}(\vec{r}) + \int d^3r' G(\vec{r}, \vec{r}', \lambda_i + \omega) V_{SCF}(\vec{r}', -\omega)\psi_i(\vec{r}'), \tag{4.35}$$

$$k^2 = \omega + \lambda_i. \tag{4.36}$$

Equation (4.35) is recognized as the basis of the Green's function method. This equation can be used to solve for $\psi_i^{(1,\pm)}$. Here we use the equation to discuss outgoing wave boundary conditions. The factor of $i\delta$ in the denominator of the Green's function gives outgoing waves.

The first term on the right side of (4.35) can be present only when k^2 is positive. Then there is a homogeneous solution to the inhomogeneous differential equation. The eigenfunction $\phi_{\vec{k}}$ is that of the electron in the final state, as it exits the atom. The factor of ϵ in front of this term is included to allow us to discuss the amount of the homogeneous solution. As a mathematics problem, ϵ can have any value, and (4.35) is still a solution to (4.31). The value of ϵ must be chosen by physical arguments. One possibility is to examine whether the constraint on the density provides any information:

$$\int d^3r n^{(1)}(\vec{r}, \omega) = 0. \tag{4.37}$$

Unfortunately, this equation provides no help in choosing ϵ since ψ_i and $\phi_{\vec{k}}$ are orthogonal because of angular factors.

However, it is easy to see that the homogeneous terms must be absent, i.e., ϵ must equal zero. The homogeneous term would correspond, in scattering theory, to an incoming plane wave. However, in photoionization, no electron wave is incoming. There is only an outgoing wave.

Our discussion assumes that the atomic system has spherical symmetry. The SCF potential has a spatial dependence of $V_{SCF}(\vec{r}) = v_{\text{SCF}}(r)P_1(\cos\theta)$. The energies in the final state ε_k are isotropic. The initial eigenstate is $\psi_i = R_{n\ell}(r)Y_{\ell m}(\theta,\phi)$. The electron eigenfunctions for the continuum and outgoing states have the form

$$\phi_{\vec{k}}(\vec{r}) = \sum_{\ell}(2\ell+1)(-i)^{\ell}u_{\ell}(kr)P_{\ell}(\hat{k}\cdot\hat{r}), \tag{4.38}$$

$$\psi_i^{(1,-)}(\vec{r}) = \sum_{\ell'=|\ell-1|}^{\ell+1} C(10;\ell m|\ell' m)R_{\ell'}^{(1,-)}(kr)Y_{\ell' m}(\theta,\phi). \tag{4.39}$$

The perturbed eigenfunction $\psi_i^{(1,-)}$ has the same form as (3.57). Here it is given in the form suitable for outgoing waves. Only $\psi_i^{(1,-)}$ has outgoing waves, while $\psi_i^{(1,+)}$ has bound solutions. The radial function $u_\ell(kr)$ is the solution to the homogeneous equation which is regular at the origin:

$$\left[-\frac{\partial^2}{\partial r^2} + \frac{\ell(\ell+1)}{r^2} + V(r) - k^2\right](ru_\ell(kr)) = 0. \tag{4.40}$$

At large distances from the atom, this eigenfunction has the asymptotic limit

$$\lim_{kr\to\infty} u_\ell(kr) \to \frac{1}{kr}\sin\left(kr + \delta_\ell(k) - \frac{\ell\pi}{2}\right). \tag{4.41}$$

The homogeneous equation is second order, and has two solutions. The second solution is called $v_\ell(kr)$. It has the asymptotic limit of $-\cos(kr + \delta_\ell - \frac{\ell\pi}{2})/(kr)$ and is irregular at the origin. The outgoing wave solution is (see Schiff, 1968)

$$w_\ell(kr) = u_\ell(kr) + iv_\ell(kr). \tag{4.42}$$

The radial part of the Green's function is obtained after averaging over angles. It is given in terms of these functions as

$$G_\ell(r,r';k) = \frac{k}{4\pi}u_\ell(kr_<)w_\ell(kr_>), \tag{4.43}$$

where $r_<$ and $r_>$ are the lesser and greater of (r, r'), respectively. In the Green's function method, the solution to the inhomogeneous differential equation for outgoing waves is

$$\begin{aligned} R_{\ell'}^{(1,-)}(kr) &= kw_{\ell'}(kr)\int_0^r r'^2 dr' u_{\ell'}(kr')v_{\rm SCF}(r', -\omega)R_{n\ell}(r') \\ &\quad + ku_{\ell'}(kr)\int_r^\infty r'^2 dr' w_{\ell'}(kr')v_{\rm SCF}(r', -\omega)R_{n\ell}(r'). \end{aligned} \tag{4.44}$$

At large distance there is the asymptotic result

$$\lim_{kr\to\infty} R_{\ell'}^{(1,-)} = Dw_{\ell'}(kr), \tag{4.45}$$

$$D = k\int_0^\infty r^2 dr u_{\ell'}(kr)v_{\rm SCF}(r, -\omega)R_{n\ell}(r). \tag{4.46}$$

These equations were solved by Zangwill and Soven (1980) in the first TDLDA calculation of photoionization. The next subsection describes how these equations are solved in the modified Sternheimer method.

4.2.2 Numerical Methods

The modified Sternheimer method can be used to calculate the outgoing eigenfunctions. Section 3.3 described the method appropriate for localized states. The procedure is simpler for outgoing states. Neither a match point nor an inward iteration is required. Instead, one just iterates outward from the origin. For example, in order to find the solution to the homogeneous equation, define $g(kr) = kru_\ell(kr)$. Use (3.43) with $T(r) = L(r)g(r)$ to iterate outward from $r = 0$ with the initial conditions $T_0 = 0, T_1 = c$. If c is chosen correctly, one has an eigenfunction of exactly the right shape, which is properly normalized. If c is chosen incorrectly, one has an eigenfunction of exactly the right shape, which is normalized incorrectly. The factor of c determines the scale factor of the eigenfunction, since the value at each subsequent point is proportional to c. However, since c is initially unknown, choose an initial value c_0 arbitrarily. Iterate to large value of r, where $V(r) \to 0$. Then the numerical solution should be a linear combination of the two plane wave solutions:

$$\lim_{kr\to\infty} g(kr) = C_g kr\cos(\delta)[j_\ell(kr) - \tan(\delta)\eta_\ell(kr)], \tag{4.47}$$

$$= C_g \sin\left(kr + \delta - \frac{\ell\pi}{2}\right). \tag{4.48}$$

The symbols $j_\ell(z)$ and $\eta_\ell(z)$ are spherical Bessel and Neumann functions, respectively. The asymptotic solution is given in terms of two constants: the amplitude C_g and the phase shift δ. The second line above has the asymptotic form given in the usual way as a sine wave shifted by phase factors which include the phase shift. The first line above is more convenient for numerical evaluation. This form is valid at intermediate values of r where the core potential $V(r)$ is zero, but the centrifugal potential $\ell(\ell+1)/r^2$ is appreciable. The two constants are obtained by fitting the numerical solution at two different points to the analytical formula (4.47). The proper normalization for $g(kr)$ is to have it be a sine wave of unit amplitude at large values of r. Then the choice for the constant T_1 is $c = c_0/C_g$.

The above procedure calculates the eigenfunction which is a solution to the homogeneous equation. The method is now described of calculating the solution $X(r) = krR^{(1,-)}(kr)$ to the inhomogeneous equation, with outgoing wave boundary conditions. We start with the calculation in LDA, i.e., the external potential z is used instead of V_{SCF}. In this case all of the radial functions $R^{(1,\pm)}(r)$ are real except those states where the electron is in an outgoing wave. Only the outgoing waves have complex eigenfunctions which contribute to $\sigma(\omega)$. The method is a simple variation on the method suggested by Senatore and Subbaswamy which was described in Section 3.3:

1. Define $X = ag + f$. The constant a and the functions $g(r)$ and $f(r)$ are determine in following fashion:

2. The function $g(r)$ is the solution to the homogeneous equation. It is found by solving the equation $g'' = Ag$ by iterating outwards from the origin with initial conditions that $g_0 = 0, g_1 = 1$. Outside of the atom it approaches the asymptotic form of

$$\lim_{r>>1} g(r) \rightarrow C_g \sin\left(kr + \delta - \frac{\ell\pi}{2}\right), \tag{4.49}$$

where the two constants C_g and δ are obtained by fitting the numerical solution at two points to this analytical form.

3. The function $f(r)$ is found by iterating the inhomogeneous equation $f'' = Af + B$ outward from the origin with the initial conditions $f_0 = 0, f_1 = 0$. For outgoing waves $B = kr\, v_{\text{SCF}}\, R_{n\ell}$. Outside of the atom it approaches the asymptotic form

$$\lim_{r>>1} f(r) \rightarrow C_f \sin\left(kr + \delta' - \frac{\ell\pi}{2}\right). \tag{4.50}$$

This asymptotic form is correct for $f(r)$ since the inhomogeneous term $B(r)$ vanishes at large r, and the solution for f becomes a combination of plane waves. The two constants C_f and δ' are determined by matching the numerical solution at two points to this analytical formula.

4. The constant a, and the function $X(r)$ are now given by

$$a = -\frac{C_f}{C_g} e^{i\delta - i\delta'}, \tag{4.51}$$

$$X(r) = f(r) - \frac{C_f}{C_g} e^{i\delta - i\delta'} g(r). \tag{4.52}$$

In order to see why this works, take the limit that r becomes large. Then use the asymptotic formulas for $g(r)$ and $f(r)$. The result is a pure outgoing wave, since the terms with $\exp(-ikr)$ cancel:

$$\begin{aligned} \lim_{r>>1} X(r) &= C_f \left[\sin\left(kr + \delta' - \frac{\ell\pi}{2}\right) - e^{i\delta - i\delta'} \sin\left(kr + \delta - \frac{\ell\pi}{2}\right)\right], \\ &= D e^{ikr + i\delta - i\frac{\ell\pi}{2}}, \qquad (4.53) \\ D &= C_f \sin(\delta' - \delta). \qquad (4.54) \end{aligned}$$

The function $X(r)$ is regular at the origin, and is an outgoing wave as r becomes large. It obeys the inhomogeneous differential equation. It is the solution we need for photoionization. Since $X(r)$ is imaginary, it contributes to the photoionization cross section. Equation (4.22) can be rewritten as

$$\sigma(\omega) = 4\pi\alpha_f \hbar\omega \int d^3r z \Im\{n^{(1)}(\vec{r},\omega)\}, \tag{4.55}$$

where α_f is the fine structure constant. The imaginary part of $X(r)$ contributes to the imaginary part of the perturbed density $n^{(1)}$.

The coefficient D is given in two different equations: (4.46) and (4.54). The theory is consistent only if these two expressions give the same numerical result. They do, and this identity provides a check on the numerical accuracy of the computer codes.

The above procedure is correct for calculating the photoionization cross section in LDA, where the external potential z is used instead of V_{SCF}. The procedure is different when the full self-consistent field is used for the external potential in (4.31). Since $n^{(1)}$ is now complex, so is V_{SCF}. Now every function $R^{(1,\pm)}(r)$ is complex, and contributes to $\sigma(\omega)$. The numerical procedure must be altered slightly when $B(r) \equiv kr\, v_{\text{SCF}}\, R_{n\ell}(r)$ is complex:

1. When $\Lambda_i = \lambda_i \pm \omega < 0$ the eigenfunctions $R^{(1,\pm)}(r)$ decay to zero outside of the atom. They are calculated using the procedure described in Section 3.3. In (3.78)–(3.79), the functions $f(r)$ and constants (a, b) are now complex. The function $g(r)$ is real since $B(r)$ is not used in computing $g(r)$. The only change in (3.78)–(3.79) is that these equations are complex. However, they are still solved in the same way.

2. When $\Lambda_i > 0$ the eigenfunctions are outgoing waves. The procedure described in (4.50)–(4.54) is altered slightly. The function $g(r)$ is unaffected. It is still real, since this function does not involve $B(r)$. The function $f(r)$ is complex, and has a different asymptotic form than (4.50):

$$\lim_{r>>1} f(r) \to C_{f_1} \sin\left(kr + \delta_1 - \frac{\ell\pi}{2}\right) + iC_{f_2} \sin\left(kr + \delta_2 - \frac{\ell\pi}{2}\right). \tag{4.56}$$

The real and imaginary parts of f have separate phase shifts (δ_1, δ_2). The coefficient a in (4.51) is changed to

$$a = -\frac{e^{i\delta}}{C_g}[C_{f_1}e^{-i\delta_1} + iC_{f_2}e^{-i\delta_2}], \tag{4.57}$$

$$X(r) = ag(r) + f(r), \tag{4.58}$$

$$D = C_{f_1}\sin(\delta_1 - \delta) + iC_{f_2}\sin(\delta_2 - \delta). \tag{4.59}$$

In computing the photoionization with the self-consistent potential, all of the eigenfunctions become complex. However, the numerical methods are the same, except that one uses complex numbers on the computer.

4.2.3 Numerical Results

Zangwill and Soven (1980) reported photoionization calculations for several atoms, including the rare gas atoms neon, argon, krypton, and xenon. For neon the results are only slightly changed between LDA and TDLDA. That is, using the self-consistent potential makes only a 20-30% change in the theoretical spectra. Neon has rather few electrons, so screening does not alter the results very much. However, for the other rare gas atoms, the difference between LDA and TDLDA is dramatic.

Figure 4.4 shows our calculated photoionization spectra for the $n = 3$ shells of argon. The dashed line is the LDA prediction, which is calculated using z instead of V_{SCF}. The solid line is the prediction of TDLDA, where

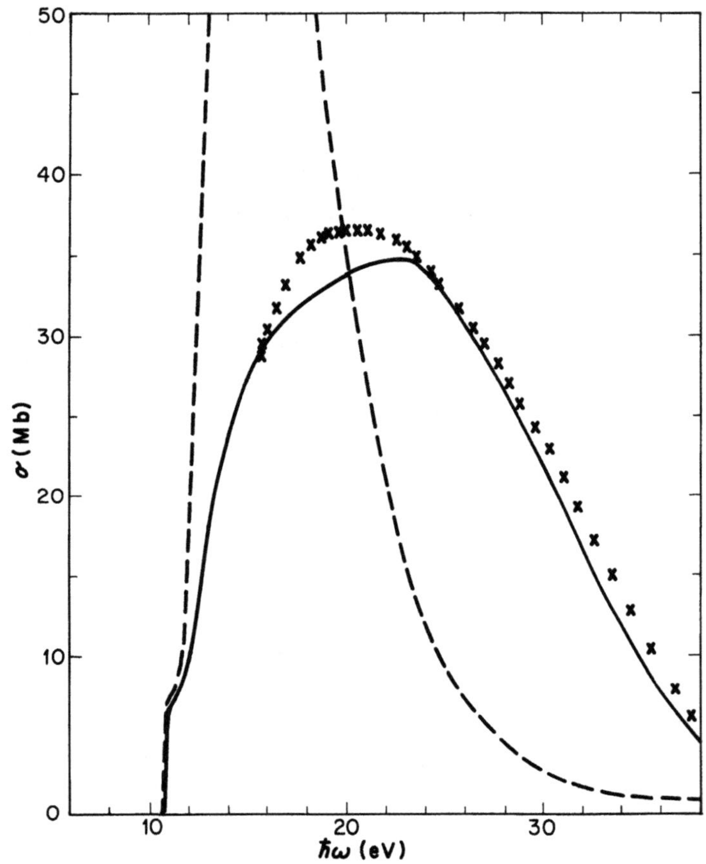

Figure 4.4: Photoionization cross section of the $n = 3$ shell of atomic argon. The dashed line is the theoretical spectrum calculated using LDA. The solid line is the theoretical spectrum calculated using TDLDA. The crosses are experimental points from Marr and West (1976)

the potential V_{SCF} is iterated until self-consistency is obtained. The crosses are experimental points from Marr and West (1976). The theoretical spectrum has a threshold of $\hbar\omega_T = 10.7$ eV which is the value of λ_{3p} given in Table 3.2. The experimental threshold is at 15.8 eV. It was remarked earlier that TDLDA gives the wrong threshold frequency.

The LDA prediction, given by the dashed line, goes up to a maximum value of 90 Mb. That is very much larger than the experimental maximum, which is about 35 Mb. There is very good agreement between the TDLDA prediction and the experimental points. They only disagree in the placement of the threshold frequency. The f-sum rule can be written as (see, e.g., Mahan, 1990)

$$\int_0^\infty \sigma(\omega)d\omega = \frac{2\pi^2 Ze^2}{mc}, \tag{4.60}$$

where Z is the number of electrons in the atom. The area under the photoionization cross section is a constant. The curves for LDA and TDLDA have the same area. In the latter case the absorption is moved out to higher frequencies. Similarly, in comparing TDLDA with the experimental spectra, the higher threshold for the experimental points means either that the experimental spectra must be larger at higher frequencies, or that the predicted spectrum below the experimental threshold accounts for the bound to bound transitions. We will return to this point in the next subsection.

The results for krypton and xenon are similar to those for argon. The $n = 4$ shell photoionization spectra of krypton is shown in Figure 4.5. The theoretical spectra calculated using TDLDA agree well with the experimental points, and are very different from the spectra calculated in LDA.

The difference between LDA and TDLDA comes from the difference between using the external potential of z compared to V_{SCF}. The self-consistent potential includes the phenomenon of screening. The ac electric field is screened by the electrons in the atoms. This screening is big, and is frequency-dependent. The TDLDA formalism provides a method of including this dynamic phenomenon in the calculation of the photoionization spectra.

Atomic theorists have invented many ways to deal with dynamic screening. Pioneering work was done by Brandt and Lundqvist (1963), and by Amusia (1963). They proposed solving (2.119) using an atomic form of the *random phase approximation* (RPA):

$$\chi(\vec{r},\vec{r}';\omega) = \chi_{HF}(\vec{r},\vec{r}';\omega) + \int d^3r_1 \chi_{HF}(\vec{r},\vec{r}_1;\omega)\chi(\vec{r}_1,\vec{r}';\omega). \tag{4.61}$$

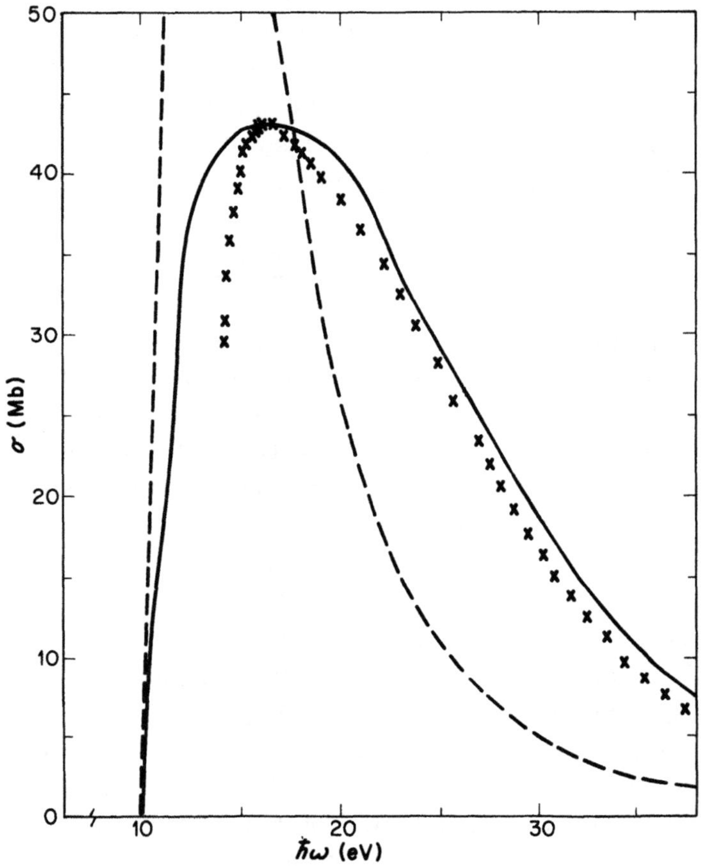

Figure 4.5: Photoionization spectra of the 4*s* and 4*p* shells of atomic krypton. Dashed line is theoretical prediction using LDA. Solid line is the theoretical prediction using TDLDA. Crosses are experimental points from Marr and West (1976)

The Hartree-Fock susceptibility χ_{HF} is identical in form to (2.121) except one uses Hartree-Fock orbitals instead of Kohn-Sham eigenfunctions. This RPA result improved the agreement between theory and experiment, but they still disagreed quite a bit. Atomic theorists then constructed modifications of RPA. Some of them, and their acronyms, are:

- RPAE: RPA with exchange. The exchange processes are added to χ_{HF}. This theory is equivalent to TDHF: time-dependent Hartree-Fock.
- SRPA: Simplified RPA with exchange. The exchange graphs are simplified, to permit easier numerical solutions. This method is easier, but still not great when compared with experiment.
- LDRPA: RPA using Kohn-Sham eigenfunctions instead of Hartree-Fock eigenfunctions. This method is similar to TDLDA in its accuracy, but is computationally more complex.

Wendin (1981, 1986) has reviewed many of these methods, and their application to photoionization and related experiments such as photoemission. TDLDA is the most accurate theory, and is the easiest numerically.

4.2.4 Self-Interaction Correction

In the previous section it was shown that TDLDA and experimental results agree for photon energies above the physical threshold. Zangwill and Soven (1980) suggested that the TDLDA absorption cross section predicted below the experimental threshold accounts for the bound to bound transitions into the autoionizing states, which are not given in TDLDA because of the incorrect behavior of the potential for large distances. This would satisfy the f-sum rule. As already discussed, performing self-interaction correction improves the long range behavior of the potential. Hence, it would be of interest to perform photoionization cross section calculations with SIC. Senatore and Subbaswamy (1990) have performed the PSIC calculation of the cross section using the modified Sternheimer method. They used the Perdew-Zunger scheme for SIC. Recall, from Section 3.5, that PSIC means that the correction is done only in the ground state. Their result for Ne is shown in Figure 4.6, along with the simple TDLDA result. Experimental data of Marr and West (1976) are shown as crosses. One sees that the threshold predicted in PSIC is in remarkably good agreement with experiment. There are several well separated bound to bound transitions descernible in the region below the threshold (not shown in the figure); these, of course, show up as

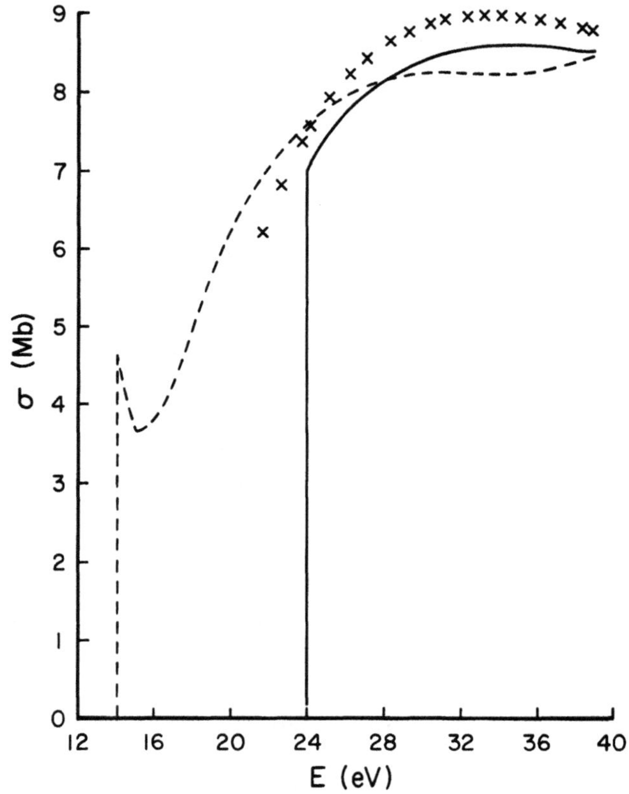

Figure 4.6: Photoionization cross section near the $2p$ threshold of atomic neon. The dashed line is the theoretical spectrum calculated using TDLDA without SIC. The solid line is calculated with SIC in the ground state, i.e., the PSIC scheme. Crosses are experimental data (Marr and West, 1976).

delta functions since no life-time effects have been put in. The TDLDA and PSIC-TDLDA cross sections are in excellent agreement in the region above threshold. Thus, it seems likely that the conjecture of Zangwill and Soven (1985) that the oscillator strength in simple TDLDA in the region below the physical threshold reflects the autoionizing resonances is correct.

4.3 $\alpha(iu)$: van der Waals Forces

4.3.1 $\alpha(iu)$

The previous section discussed using LDA for the calculation of the frequency-dependent conductivity $\alpha_d(\omega)$. Its imaginary part gave the optical absorption. This section discusses the calculation of the conductivity evaluated at a frequency which is entirely imaginary. This function is called $\alpha_d(iu)$ where the variable u is real. Sometimes we write it as $\alpha_{d,i}(iu)$ where the additional subscript i labels the ion or atom. The polarizability at an imaginary frequency is a theoretical quantity with no direct experimental counterpart. However, the function $\alpha_d(iu)$ is useful for calculating a variety of parameters which can be measured. Many of these parameters are related to van der Waals forces.

The polarizability $\alpha_d(iu)$ is a real function. In terms of oscillator strengths $f_{\beta j}$ and excitation energies $\epsilon_{\beta j} = \lambda_\beta - \lambda_j$, it is defined as

$$\alpha_d(iu) = \frac{e^2}{m}\sum_{j\beta}\frac{f_{\beta j}}{\epsilon_{\beta j}^2 + u^2}, \tag{4.62}$$

$$f_{\beta j} = 2m\epsilon_{\beta j}| < \beta|V_{SCF,1}|j > |^2/\hbar^2. \tag{4.63}$$

The electron states j are occupied and the states β are unoccupied. The function $\alpha_d(iu)$ falls off smoothly as a function of u.

The first calculations of $\alpha_d(iu)$ were done according to this formula. The first step was to calculate or measure the oscillator strengths and excitation energies. Then one evaluated the above formula for different values of u. This method of evaluation is both cumbersome and wrong. It is wrong since the self-consistent potential $V_{SCF,1}$ depends upon frequency. This feature is ignored in the above evaluation. It is cumbersome since there are many terms in the above summation. Also, the integral must be done over the excitation states β in the continuum. Since $\alpha_d(iu)$ is a very useful quantity, many calculations of this type have been one: e.g., Rehr, Zaremba, and Kohn (1975) who studied the metals copper, silver, and gold. Leonard and Barker (1975) studied the rare gases.

The first good evaluation of $\alpha_d(iu)$ was by Ahlberg and Goscinski (1975). They used a self-consistent Xα technique. It is similar to the method described below, but lacks the correlation potential. Their results are slightly less accurate then the values found using TDLDA, but were a big improvement over previous evaluations. Their result for Ne is shown in Figure 4.7.

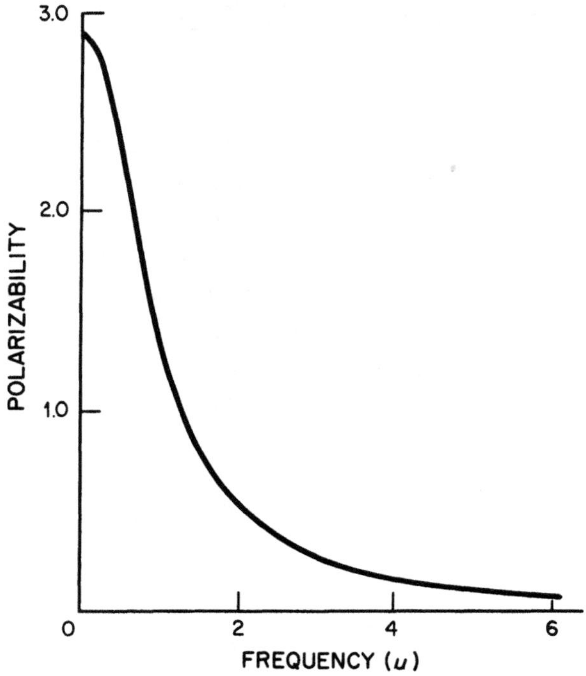

Figure 4.7: $\alpha_d(iu)$ for neon as calculated by Ahlberg and Goscinski (1975) using the Xα method.

The present method of calculating $\alpha_d(iu)$ is easier than evaluating (4.62). One uses the standard TDLDA method with the eigenvalue of $\lambda_j + iu$. Nonzero frequency results were treated in the prior section. The polarizability at finite frequency is obtained by evaluating the change in the wave function $\psi_j^{(1)}$ using eigenvalues $\lambda_j \pm \omega$. The same technique is used for complex frequencies. Since only the real part of $\psi_j^{(1)}$ is needed, it does not matter whether we use $\lambda_j + iu$ or $\lambda_j - iu$. The equation to be solved is

$$[H_0 - \lambda_j - iu]\psi_j^{(1)}(\vec{r}) = -V_{SCF,1}\psi_j(\vec{r}), \tag{4.64}$$

$$n^{(1)}(\vec{r}) = 2\Re\{\sum_j \psi_j^* \psi_j^{(1)}\}. \tag{4.65}$$

The first-order change in density $n^{(1)}$ is still a real function. The self-consistent potential is still evaluated according to (4.10) and is also real. The

Table 4.8: $\alpha_d(iu)$ in units of a_0^3. u in units of Rydbergs.

u	Ne	Na^+	Ar	K^+	Kr	Rb^+
0.0	3.000	1.073	11.811	5.706	17.710	9.422
0.5	2.765	1.049	10.074	5.329	14.508	8.580
1.0	2.313	0.991	7.290	4.485	9.900	6.841
1.5	1.888	0.912	5.168	3.580	6.723	5.178
2.0	1.547	0.827	3.745	2.815	4.723	3.900
2.5	1.284	0.744	2.796	2.222	3.452	2.979
3.0	1.079	0.667	2.150	1.776	2.608	2.320
3.5	0.918	0.599	1.696	1.442	2.029	1.843
4.0	0.791	0.538	1.369	1.188	1.616	1.492
4.5	0.687	0.485	1.126	0.993	1.315	1.228
5.0	0.603	0.438	0.942	0.841	1.089	1.026
5.5	0.533	0.398	0.799	0.721	0.915	0.868
6.0	0.474	0.362	0.686	0.624	0.779	0.743
6.5	0.425	0.330	0.595	0.546	0.671	0.623
7.0	0.382	0.303	0.521	0.481	0.583	0.561
7.5	0.346	0.278	0.461	0.427	0.512	0.494
8.0	0.315	0.256	0.410	0.381	0.452	0.438
8.5	0.288	0.237	0.367	0.343	0.403	0.390
9.0	0.264	0.220	0.330	0.310	0.361	0.350
9.5	0.242	0.204	0.299	0.282	0.325	0.316
10.0	0.224	0.190	0.272	0.257	0.294	0.287

only function which is complex is the first-order eigenfunction $\psi_j^{(1)}$. When compared to the evaluation of the static polarizability, the only change in the procedures is to make the eigenvalue a complex quantity when evaluating the first-order change in the eigenfunction $\psi_j^{(1)}$. The latter function is still evaluated according to the methods given in Section 3.4. The functions X, f, g are all complex, but they are found by the same matching technique. Table 4.8 shows $\alpha_d(iu)$ as a function of u for three atoms and three ions. Mahan (1982) actually calculated the values for fourteen different atoms and ions. The values fall off smoothly with increasing u, and fall off at a different rate for each case. At very large values of u the polarizability has the limiting value of

$$\lim_{u \to \infty} \alpha_d(iu) \to \frac{e^2}{mu^2} \sum_{\beta j} f_{\beta j} = \frac{Z' e^2}{mu^2}, \tag{4.66}$$

where we have used the fact that the f-sum rule equals the number of electrons Z' in the atom or ion. The numerical values in Table 4.8 do not approach this limit at $u = 10$. The limiting formula is valid only when $u >> \epsilon_{\beta j}$ which is far from being satisfied at $u = 10$.

4.3.2 van der Waals Interaction

The van der Waals interaction is the long range potential between two neutral atoms or molecules. It is an important interaction in atomic and chemical physics. Let R denote the separation of the two atoms or molecules. This distance is assumed to be large enough that the atoms no longer touch—their electronic orbitals are well separated. Then the potential energy can be expressed as (see Bell and Zucker, 1976)

$$V(R) = -\frac{C_6}{R^6} - \frac{C_8}{R^8} - \frac{C_{10}}{R^{10}} - \ldots. \tag{4.67}$$

The parameter C_6 is called the van der Waals coefficient. Its value depends upon the identity of the two atoms or molecules. There is a different coefficient for each pair of atoms or molecules. Most molecules are not spherically symmetric. Then the coefficient C_6 depends upon the relative orientation of the two molecules. Equation (4.67) is the nonrelativistic version of the long-range interaction. At very large separations relativistic corrections change the R^{-6} power law into an R^{-7} law. However, the interaction is very weak at those long distances, so the inverse seventh power is rarely observed.

The van der Waals interaction can be expressed in a simple formula by using the polarizability $\boldsymbol{\alpha}_d(iu)$. Boldface type is used for the polarizability to indicate it is a second-rank tensor. The tensor feature is important for molecules which are not spherically symmetric, or for atoms with partially filled electron shells. The other tensor in the formula is the dipole-dipole interaction $\boldsymbol{\phi}$. The symbol 'Tr' denotes trace. The first term of (4.67) is:

$$V_{vdW}(R) = -\int_0^\infty \frac{du}{2\pi} Tr\{\boldsymbol{\alpha}_{d,i}(iu) \cdot \boldsymbol{\phi}(\boldsymbol{R}) \cdot \boldsymbol{\alpha}_{d,j}(iu) \cdot \boldsymbol{\phi}(\boldsymbol{R})\ \}, \tag{4.68}$$

$$\phi_{\mu\nu}(\vec{R}) = \frac{\delta_{\mu\nu}}{R^3} - \frac{3R_\mu R_\nu}{R^5}. \tag{4.69}$$

Table 4.9: C_6 in units of $e^2a_0^5$.

	Ne	Na^+	Ar	K^+	Kr	Rb^+
Ne	7.14	3.36	21.4	13.0	29.9	19.6
Na^+		1.70	9.45	6.00	12.8	8.84
Ar			68.2	39.9	95.3	61.2
K^+				24.1	55.1	36.4
Kr					133.6	85.0
Rb^+						55.4

The angular dependence is rather complicated when the molecules are asymmetric.

This formula simplifies for the case when the two atoms or molecules i and j are both spherically symmetric. Then the polarizabilities are each a scalar function $\alpha_d(iu)$ times the unit tensor. The interaction has the simple form of the first term in (4.67) and the coefficient C_6 does not depend upon the orientation of the atoms or molecules

$$C_{6,ij} = \frac{3}{\pi}\int_0^\infty du\alpha_{d,i}(iu)\alpha_{d,j}(iu). \tag{4.70}$$

This formula is simple to evaluate after one has calculated the polarizability $\alpha_d(iu)$ for a number of atoms or ions. The van der Waals coefficient between two atoms is then obtained by a simple convolution integral of their two polarizabilities. This calculation was first done by Mahan (1982).

The numerical results for $\alpha_d(iu)$ are used in equation (4.70). Doing the integral over u produces a value for C_6. The integral is done for each pair of ions, which produces a six-by-six array of values, which are shown in Table 4.9. Here the dimensional units are $e^2a_0^5$ which is traditional for this constant.

London (1930) suggested a simple and accurate way to understand the van der Waals coefficient. He suggested that the frequency dependence of the polarizability could be described by a parameter E_i:

$$\alpha_{d,i}(iu) \approx \frac{\alpha_{d,i}(0)}{1+u^2/E_i^2}, \tag{4.71}$$

$$C_{6,ij} \approx \frac{3}{2}\alpha_{d,i}(0)\alpha_{d,j}(0)\frac{E_iE_j}{E_i+E_j}. \tag{4.72}$$

The usefulness of the formula depends upon finding the best value for the characteristic energy E_i for each atom or ion. A simple and unique way to

Table 4.10: Theoretical values of $\alpha_{d,i}(0)$ (in units of a_0^3), $C_{6,ii}$ (in units of $e^2a_0^5$), and E_i (in units of $e^2/2a_0$).

Ion	$\alpha_{d,i}(0)$	$C_{6,ii}$	E_i
Ne	3.00	7.14	2.12
Na^+	1.07	1.70	3.93
Ar	11.8	68.2	1.30
K^+	5.71	24.1	1.97
Kr	17.7	134.	1.14
Rb^+	9.42	55.4	1.66

determine this energy is from the coefficient between two identical atoms or ions,

$$E_i = \frac{4}{3}\frac{C_{6,ii}}{\alpha_{d,i}(0)^2}. \tag{4.73}$$

This formula gives the results in Hartrees since the units of C_6 are $e^2a_0^5$. Multiplying by 2 gives the values in Rydbergs. Values obtained this way are shown in Table 4.10.

There is an easy way to test the London formula (4.72). One can calculate $C_{6,ij}$ for $i \neq j$ using this formula with the values of $\alpha_{d,i}(0)$ and E_i shown in Table 4.10. Then compare these results with the values in Table 4.9 obtained by doing the full integral in (4.70). The values calculated by the two methods agree to within 1% for all cases. The London formula works quite well. For this reason we shall use it again to calculate other quantities.

So far all of the numerical values presented in this section have been theoretical ones calculated using TDLDA. We now turn to comparison with experimental values. Good values for C_6 are available for the rare gas atoms. They are obtained by atom-atom scattering using atomic beams. The long-range force determines the small angle scattering. The scattering does show an R^{-6} behavior, and the measurement determines the value of C_6. Less accurate values are found from interatomic potentials as measured by chemical reaction constants. Table 4.11 shows values of E_i as determined from (4.73) using experimental values for $\alpha_{d,i}(0)$ and C_6. These experimental values for C_6 and E_i should be compared with the theoretical values in Table 4.10. For argon and krypton, the theoretical values of C_6 agree within 1% of the experimental measurements. This agreement is probably a bit fortuitous, since the polarizability is about 7% high. When two ions are apart, their

Table 4.11: Experimental values of $\alpha_d(0)$, C_6, and E_i as deduced from (4.74). Units are the same as in Table 4.10.

Atom	$\alpha_{d,i}(0)^a$	C_6^b	E_i
He	1.38	1.47	2.06
Ne	2.70	6.4	2.34
Ar	11.07	65	1.41
Kr	16.74	131	1.25

[a] R.R. Teachout and R.T. Pack, Atomic Data **3**, 195 (1971)

[b] A. Dalgarno and W.D. Davison, *Advances in Atomic and Molecular Physics* **2**, ed. D.R. Bates and I. Estermann (Academic Press, 1966) p.1

interaction energy has a number of terms caused by the net valence charge Z_i on the ion. The charge on one ion causes moments to form on the other, so the interaction energy depends upon the static polarizabilities

$$V_{ij} = Z_i Z_j \frac{e^2}{R} - \frac{e^2}{2R^4}[Z_i^2 \alpha_{d,j} + Z_j^2 \alpha_{d,i}] - \frac{e^2}{2R^6}[Z_i^2 \alpha_{q,j} + Z_j^2 \alpha_{q,i}] - \cdots. \quad (4.74)$$

There is an interaction of the order R^{-6} caused by the ion charge inducing a quadrupole moment on the other atom or ion. This interaction has the same power law as the van der Waals interaction. The total coefficient of the R^{-6} term is

$$C_{6,ij} + \frac{e^2}{2}[Z_i^2 \alpha_{q,j} + Z_j^2 \alpha_{q,i}]. \quad (4.75)$$

The first term is from the van der Waals interaction, while the other two terms are from the quadrupole polarizability. The last two terms are usually smaller, but not negligible.

The method used above to calculate $\alpha_d(iu)$ and C_6 using TDLDA can be used to evaluate the other coefficients in (4.67). Dalgarno (1967) notes that one can define integrals similar to (4.70)

$$C_{ij}(M,L) = \frac{(2M+2L)!}{4(2M)!(2L)!} \frac{2}{\pi} \int_0^\infty du \alpha_{M,i}(iu) \alpha_{L,j}(iu), \quad (4.76)$$

$$C_{6,ij} = C_{ij}(1,1), \quad (4.77)$$

$$C_{8,ij} = C_{ij}(1,2) + C_{ij}(2,1), \quad (4.78)$$

$$C_{10,ij} = C_{ij}(1,3) + C_{ij}(2,2) + C_{ij}(3,1). \qquad (4.79)$$

The coefficient C_8 is from the mutual polarization of the dipole on one atom with the quadrupole on the other. The coefficient C_{10} is from the mutual polarization of the two quadrupole moments, and also from the mutual polarization of a dipole moment on one atom with an octupole moment on the other. These coefficients have been calculated by Tang, Norbeck, and Certain (1976) using London's approximation for $\alpha_{M,i}(iu)$, which is the Mth polarizability on the ith atom. Theoretical Hartree-Fock values were used for the values of the static polarizabilities with $M > 1$.

4.3.3 Axilrod-Teller Interaction

The van der Waals interaction comes from the mutual polarization of two atoms, ions, or molecules. In a similar way, Axilrod and Teller (1943) considered the interaction of three ions caused by their mutual polarization. The general formula for the potential energy of interaction is

$$V_{AT} = -\int_0^\infty \frac{du}{\pi}\, Tr\{\boldsymbol{\alpha}_{d,1}(i\boldsymbol{u}) \cdot \boldsymbol{\phi}(\boldsymbol{R}_{12}) \cdot \boldsymbol{\alpha}_{d,2}(i\boldsymbol{u}) \cdot \boldsymbol{\phi}(\boldsymbol{R}_{23}) \cdot \boldsymbol{\alpha}_{d,3}(i\boldsymbol{u}) \cdot \boldsymbol{\phi}(\boldsymbol{R}_{31})\}. \qquad (4.80)$$

The interaction has a complicated form when the molecules are anisotropic. A relatively simple formula is obtained for the Axilrod-Teller interaction between three atoms or ions with all electrons in closed shells. When the three atoms or ions are identical, the Axilrod-Teller interaction is

$$V_{AT} = -\frac{C_{AT}}{(R_{12}R_{23}R_{31})^3}[1 + 3\cos(\theta_1)\cos(\theta_2)\cos(\theta_3)], \qquad (4.81)$$

$$C_{AT} = \frac{3}{\pi}\int_0^\infty du\, [\alpha_d(iu)]^3. \qquad (4.82)$$

The three angles θ_j are the interior angles of the triangle formed by the three atoms at the vertices. The strength of the interaction is given by the Axilrod-Teller coefficient C_{AT}. It is easy to evaluate once $\alpha_{d,i}(iu)$ is determined.

The integral over du is evaluated easily by using the London approximation (4.71) for $\alpha_{d,i}(iu)$, to get

$$C_{AT} = \frac{9}{16}\alpha_{d,i}(0)^3 E_i. \qquad (4.83)$$

Table 4.12: First two columns are the Axilrod-Teller coefficient C_{AT} in units of $e^2a_0^8/2$. Third column is nearest neighbor distance R_1 in the solid at T=0 K. The last two columns show the magnitude of the Axilrod-Teller and van der Waals potentials in Rydbergs.

Atom	C_{AT}		R_1	C_{AT}/R_1^9	C_6/R_1^6
	(a)	(b)	(Å)	(10^{-5}Ry)	(10^{-3}Ry)
Ne	25.9	23.6	5.965	0.27	0.28
Ar	1088	1042	7.097	2.4	1.0
Kr	3290	3120	7.544	4.2	1.4

[a] Present results

[b] P.W. Langhoff and M. Karplus, J. Chem. Phys. **53**, 233 (1970)

In Table 4.12 values for C_{AT} evaluated from this formula and the experimentally derived values from Table 4.11 are shown. Column (a) are the present results, while column (b) are the results of a similar calculation by Langhoff and Karplus (1970). Their calculation was similar in that they used the static polarizability and van der Waals coefficient to find E_i, and then used these numbers to find C_{AT}. Their values are slightly different, which reflects the small change in the experimental parameters over twenty years. Similar results were also found by Doren (1974).

An estimate of the magnitude of this potential is obtained by calculating C_{AT}/R_1^9 where R_1 is the first neighbor distance in the solid. These values are shown in Table 4.12, and are compared to the magnitude of the van der Waals interaction. The latter are 2-3 orders of magnitude larger. The Axilrod-Teller interaction can usually be neglected in metals and semiconductors. However, in solids bound by van der Waals potentials it may need to be included in the calculation of the ground state energy. The interaction energy from a single Axilrod-Teller interaction is much smaller than the similar contribution from the van der Waals term. However, there are many more three-body combinations than two-body combinations. A thorough discussion for the rare gas solids is given by Bell and Zucker (1976). In solid xenon the Axilrod-Teller interaction is calculated to provide 10% of the ground state energy. The contribution is smaller in other solids.

4.3.4 Multiatom Interactions

The previous sections gave formulas (4.68) and (4.80) for the ground state energy of the dipolar interaction for two and three atoms, respectively. These formulas are generalized to a collection of N atoms. The first step is to derive the equation for the frequencies of the normal modes of oscillation in the system of N dipoles. Each induced dipole $\vec{p}_j$ has its value determined by the local field from the other dipoles,

$$\boldsymbol{p}_j = -4\pi \sum_{i \neq j} \boldsymbol{\alpha}_{d,j}(\omega) \cdot \boldsymbol{\phi}(\boldsymbol{R}_{ji}) \cdot \boldsymbol{p}_i. \tag{4.84}$$

The above set of equations have a dimension of $3N$. The normal modes are obtained by setting the determinant equal to zero

$$D_N(\omega) = \det |\delta_{ij}\delta_{\mu\nu} + 4\pi(\alpha_{d,j}(\omega))_{\mu\lambda}\phi_{\lambda\nu}(\vec{R}_{ji})| = 0. \tag{4.85}$$

The long-range attractive interation between the atoms is due to the zero point motion of these normal modes. This interaction is, in the dipole approximation, given exactly (Mahan, 1965) by

$$V = \frac{1}{2\pi} \int_0^\infty du \ln D_N(iu). \tag{4.86}$$

Both the van der Waals interaction, and the Axilrod-Teller interaction, are obtained by expanding this expression. The expansion utilizes the formula:

$$\ln\{\det |\boldsymbol{I} + \boldsymbol{W}|\} = \mathrm{Tr}\,[\, \boldsymbol{W} - \tfrac{1}{2}\boldsymbol{W} \cdot \boldsymbol{W} + \tfrac{1}{3}\boldsymbol{W} \cdot \boldsymbol{W} \cdot \boldsymbol{W} \ldots], \tag{4.87}$$

where $\boldsymbol{W} = 4\pi\boldsymbol{\alpha} \cdot \boldsymbol{\phi}$. The van der Waals and Axilrod-Teller interaction come from the second and third terms in this expansion. More terms can be calculated as needed.

Multiatom polarizations become smaller as the number of atoms increases above 2. This conclusion is comforting, since it shows that perturbation theory is converging.

4.4 Hyperpolarizability

Even though nonlinear optical effects have been known since the latter half of the nineteenth century, they received prominence since the invention of the laser. The practical interest in crystals suitable for frequency doubling and mixing led to the measurements of nonlinear susceptibilities of various materials. The early measurements were not very accurate. Also, theoretical calculations were confined mostly to the lightest of atoms, most notably He. In the seventies and eighties there has been a revival of interest in the measurement of nonlinear susceptibilities because the ready availability of high powered lasers and detectors has led to much better accuracy. This also rekindled the interest in theoretical calculations.

Since nonlinear susceptibilities are not as commonly known as their linear counterparts, we begin with some preliminaries. So far, we have introduced two related quantities: the hyperpolarizability γ of (1.1), and the susceptibility $\chi^{(3)}$ of (2.45). The former is a microscopic quantity characteristic of the atoms or ions, while the latter is a macroscopic quantity, characteristic of the medium as a whole. Just as the linear polarizability α is related to the macroscopic linear susceptibility $\chi^{(1)}$ via the Clausius-Mossotti relationship, the nonlinear susceptibilities and polarizabilities have a similar relationship. Let $\vec{P}$ denote the polarization, i.e., the dipole moment density of the medium, and $\vec{p}$ the dipole moment of the atoms in the medium. We are assuming a monatomic system for simplicity, and considering an isotropic medium. Then, one can write

$$\vec{D} = \vec{\mathcal{E}} + 4\pi\vec{P}, \tag{4.88}$$
$$\vec{P} = \chi^{(1)}\vec{\mathcal{E}} + \vec{P}^{NL}, \tag{4.89}$$

where $\vec{P}^{NL}$ is the nonlinear part of the polarization, and $\vec{\mathcal{E}}$ is the macroscopic field, related under the usual approximations (see, e.g., Ashcroft and Mermin, 1976) to the local field $\vec{\mathcal{E}}_{loc}$ via

$$\vec{\mathcal{E}}_{loc} = \vec{\mathcal{E}} + \frac{4\pi}{3}\vec{P}. \tag{4.90}$$

Let Ω_0 be the volume per atom in the medium, i.e., the number density is Ω_0^{-1}. Then,

$$\vec{P} = \Omega_0^{-1}(\alpha\vec{\mathcal{E}}_{loc} + \vec{p}^{NL}), \tag{4.91}$$

where α is the linear polarizability of the atoms. Substituting from (4.90) it is readily seen that

$$\vec{P} = \frac{\Omega_0^{-1}(\alpha\vec{\mathcal{E}} + \vec{p}^{NL})}{1 - (4\pi\alpha/3\Omega_0)}. \tag{4.92}$$

Thus, for the linear term, one gets the familiar result

$$\chi^{(1)} = \frac{\epsilon - 1}{4\pi} = \frac{\Omega_0^{-1}\alpha}{1 - (4\pi\alpha/3\Omega_0)}, \tag{4.93}$$

whence

$$\frac{\Omega_0^{-1}}{1-(4\pi\alpha/3\Omega_0)} = \frac{\epsilon+2}{3\Omega_0}. \tag{4.94}$$

Thus, finally, we get the central result (see, e.g., Bloembergen, 1965),

$$\vec{P}^{NL} = \left(\frac{\epsilon+2}{3\Omega_0}\right)\vec{p}^{NL}. \tag{4.95}$$

If one now considers the definitions, e.g.,

$$p^{NL} = \frac{\gamma}{6}\mathcal{E}_{loc}^3 + \cdots, \tag{4.96}$$

$$P^{NL} = \chi^{(3)}\mathcal{E}^3 + \cdots, \tag{4.97}$$

one gets the following relation between γ and $\chi^{(3)}$:

$$\chi^{(3)} = \frac{\gamma}{6\Omega_0}\left(\frac{\epsilon+2}{3}\right)^4. \tag{4.98}$$

A more detailed expression, valid in the phonon region for the crystalline state, will be derived in Section 5.7.

4.4.1 Closed-Shell Atoms

For a spherically symmetric atom or ion the lowest nonvanishing nonlinear susceptibility is the hyperpolarizability γ discussed above. The first attempt at extending the LDA treatment of polarizabilities to the calculation of γ was by Zangwill (1983), who applied it to rare gas atoms. However, his calculation was not completely self-consistent, and the first correct calculation was that of Subbaswamy and Mahan (1985) for He. A thorough treatment of all rare gas atoms, including the effect of self-interaction corrections was given subsequently by Senatore and Subbaswamy (1986).

In this section we confine our attention to the static case. For a spherically symmetric atom or ion, the fourth-rank tensor γ_{ijkl} has only one independent component (Buckingham, 1967):

$$\begin{aligned} \gamma_{zzzz} &= \gamma, \\ \gamma_{xxzz} &= \gamma_{xzxz} = \gamma/3. \end{aligned} \tag{4.99}$$

Most of the expressions needed for the evaluation of γ have been given earlier either explicitly or implicitly. One has (see (2.100))

$$\frac{\gamma/6}{a_0^7/e^2} = -\int d^3r\; n^{(3)}(\vec{r})\; r\cos\theta. \tag{4.100}$$

On the right-hand side, all lengths are in Bohr radii and all energies in Rydbergs. We note for future reference that

$$a_0^7/e^2 = 0.5037 \times 10^{-39}\text{esu}. \tag{4.101}$$

The third-order change in density, $n^{(3)}$ can be calculated either by the modified Sternheimer method outlined in Section 3.3, or by the Green's function method of Section 3.4. Both methods are comparable in computational effort. The most tedius part is the separation of different angular momentum components of the various quantities entering the calculation. An example is given in the Appendix to this section.

For the static case under consideration here, γ can be calculated without explicitly evaluating the third-order orbitals or density. This may be accomplished using the expression for the total energy,

$$E = E_0 - \frac{1}{2}\alpha\mathcal{E}^2 - \frac{1}{24}\gamma\mathcal{E}^4 + \cdots. \tag{4.102}$$

The appropriate expression in LDA is derived below. Recall the LDA expression for the total energy, (2.35), which is rewritten below for convenience:

$$E = \sum_j \lambda_j - \int d^3r_1 d^3r_2 \frac{n(\vec{r}_1)n(\vec{r}_2)}{\mid \vec{r}_1 - \vec{r}_2 \mid} + \int d^3r\, n(\vec{r})[\varepsilon_{xc}(n(\vec{r})) - V_{xc}(n(\vec{r}))]. \tag{4.103}$$

Now, all quantities are expanded in powers of the applied field as before. Then, one has to collect all terms of fourth-order in the applied field. The expression for $\lambda_j^{(4)}$ is deduced from (2.57). The term $< \psi_j^{(0)} \mid V^{(1)} \mid \psi_j^{(3)} >$ involving the third-order orbital is simplified by taking the inner product of the third-order equation in (2.55) with $\psi_j^{(0)}$ to get

$$\begin{aligned} < \psi_j^{(0)} \mid V^{(1)} \mid \psi_j^{(3)} > &= < \psi_j^{(1)} \mid V^{(1)} \mid \psi_j^{(2)} > \\ &+ < \psi_j^{(1)} \mid V^{(2)} - \lambda_j^{(2)} \mid \psi_j^{(1)} > + < \psi_j^{(1)} \mid V^{(3)} \mid \psi_j^{(0)} > . \end{aligned} \tag{4.104}$$

Thus, finally, one gets the following expression for the fourth-order energy, which is expressed in terms of first- and second-order Kohn-Sham orbitals, eigenvalues, and potentials:

$$\begin{aligned} E^{(4)} &= \frac{1}{2}\int d^3r \left\{ \sum_j \left[V^{(1)}(\psi_j^{(1)*}\psi_j^{(2)} + \text{c.c.}) + (V^{(2)} - \lambda_j^{(2)}) \mid \psi_j^{(1)} \mid^2 \right] \right. \\ & \left. + \frac{1}{12}\left[n^{(1)^4} v_{xc}''' + 6n^{(1)^2} n^{(2)} v_{xc}'' \right] \right\}. \end{aligned} \tag{4.105}$$

The various order potentials are defined in (2.113)-(2.115). The numerical procedure for solving the inhomogeneous Schrödinger equation for the first-order orbital has already been described in Section 3.3. The procedure for the second-order orbital is identical. The angular momentum decomposition of the equation is described in the Appendix. The first-order equation is first solved to self-consistency. The first-order orbitals, density, and potential are then used to construct the initial guess for the inhomogeneous terms in the second-order Kohn-Sham equation. Then the second-order orbital, density, and potential are found self-consistently. Convergence is obtained readily. Note that in the solution of the second-order orbital for $2s$, $3p$, etc., the orthogonality conditions resulting from (3.80) must be used, as discussed in Section 3.3.

In Table 4.13 the calculated values of the hyperpolarizability of rare gas atoms are compared with available experimental results. The calculations were performed by the modified Sternheimer scheme using the Ceperley-Alder exchange-correlation energy. The values obtained when self-consistent screening is neglected (denoted LDA) are also shown. Four sets of experimental results are shown. The values shown were obtained by extrapolation to zero frequency from optical frequency measurements. The measured dispersion in $\gamma(\omega)$ from the work of Mizrahi and Shelton (1985) in conjunction with the one resonance oscillator model of Shelton (1986) was used for this purpose. For He, the accurate variational calculation by Sitz and Yaris (1968) is quoted. In fact, most measurements are made relative to He standard. The electric-field-induced second harmonic generation (ESHG) values are quoted with the smallest uncertainties. Considering the quoted uncertainties in the other measurements (third harmonic generation (THG), four-wave mixing (FWM), and dc-Kerr), and the additional errors introduced by the extrapolation to zero frequency, the experimental values are consistent with one another.

Note first of all that the effect of screening is very significant in the hyperpolarizability. The most substantial part of the screening comes from the Coulomb term. The TDLDA overestimates γ systematically by nearly a factor of 2. This is merely a large amplification of the discrepancy found earlier for the linear polarizability. As in the case of α, we find that use of the Gunnarsson-Lundqvist exchange correlation potential lowers the TDLDA values by about 10%, leaving a substantial disagreement with expermental values. This is understandable, since the second- and third-order orbitals are diffuse, and the LDA potential has incorrect behavior in that region. The

Table 4.13: Hyperpolarizability $\gamma/6$ of rare gases (in 10^{-39} esu). LDA is without, and TDLDA is with, screening. The experimental acronyms are discussed in the text.

	He	Ne	Ar	Kr	Xe
LDA	9.08	24.49	272	644	1656
TDLDA[a]	7.40	17.7	156	332	769
ESHG[b]	3.58[f]		92±2	188± 4	404±8
THG[c]	3.5±0.4	5.9±0.6	72±7	174±17	403±40
FWM[d]	3.0±0.7		105±19	229±41	
dc Kerr[e]	4.4±0.3	8.0±0.6	91±6	202±14	489±38

[a]G. Senatore and K.R. Subbaswamy, Phys. Rev. A **34**, 3619 (1986)
[b]V. Mizrahi and D.P. Shelton, Phys. Rev. Lett. **55**, 696 (1985)
[c]H.T. Lehmeier *et al.*, Opt. Commun. **56**, 67 (1985)
[d]M. Lundeen *et al.*, J. Chem. Phys. **79**, 6301 (1983)
[e]A.D. Buckingham and D. A. Dunmur, Trans. Faraday Soc. **64**, 1776 (1968)
[f]Accurate theoretical value by P. Sitz and R. Yaris, J. Chem. Phys. **49**, 3546 (1968); this value was used as the reference value in the ESHG measurements[b]

zeroth- and third-order Kohn-Sham orbitals for the $2p$ shell in Ne are shown in Figure 4.8.

The agreement is expected to improve with SIC. But, this poses the artificial singularity problem already discussed. Senatore and Subbaswamy (1986) used the Perdew-Zunger self-interaction correction scheme only in the unperturbed state; this is the scheme we have referred to as PSIC before. In Table 4.14 we compare the results with and without PSIC. As expected, the SIC makes the atoms more rigid. The size of the correction becomes progressively smaller in going from He to Xe. This trend is to be expected on general grounds; however, the possiblility that this is an artifact of the particular SIC scheme adopted can not be ruled out. In Table 4.14 the most recent *ab initio* results of Sitter and Hurst (1972) are also shown. These were performed at the Hartree-Fock level. As may be seen, these values give roughly the same degree of agreement with experiments as the PSIC calculations (see Table 4.14).

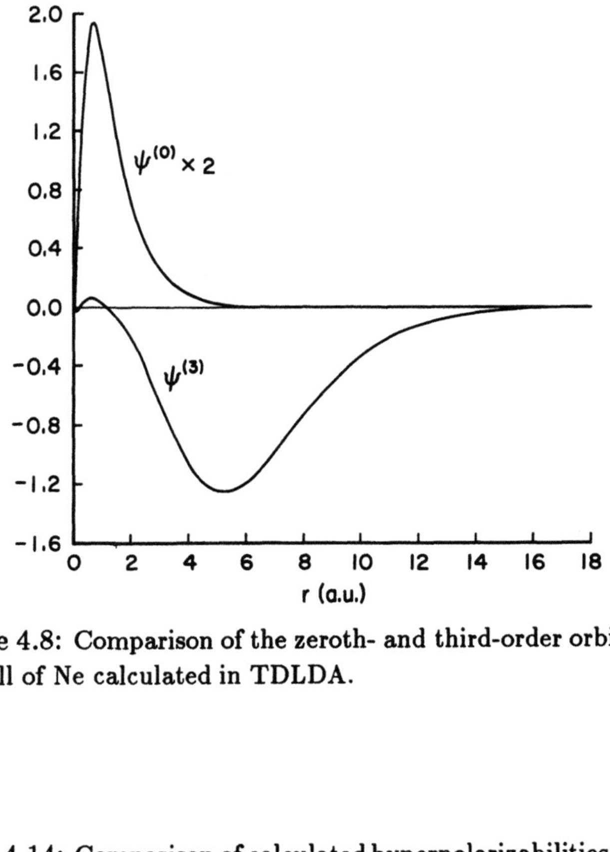

Figure 4.8: Comparison of the zeroth- and third-order orbitals for $2p$ shell of Ne calculated in TDLDA.

Table 4.14: Comparison of calculated hyperpolarizabilities for rare gases. TDLDA and PSIC are, respectively, without and with self-interaction correction in the unperturbed state. HF is Hartree-Fock.

	He	Ne	Ar	Kr	Xe
TDLDA[a]	7.40	17.7	156	332	769
PSIC[a]	2.92	7.18	112	269	701
HF[b]	3.02	3.52	84.8		

[a]G. Senatore and K.R. Subbaswamy, Phys. Rev. A **34**, 3619 (1986)
[b]R.E. Sitter and R.P. Hurst, Phys. Rev. A **5**, 5 (1972)

Appendix

In this Appendix, we outline the angular momentum decomposition of the second-order Kohn-Sham equation. More details may be found in Senatore and Subbaswamy (1986). We are interested in the solution of

$$(H_0 - \lambda_{nl}^{(0)})\psi_{nlm}^{(2)} = -V^{(1)}\psi_{nlm}^{(1)} + (\lambda_{nl}^{(2)} - V^{(2)})\psi_{nlm}^{(0)}, \tag{4.106}$$

where all quantities have been defined before. The external potential is assumed to have the form

$$V_{ext}(\vec{r}) = v_M(r)P_M(\cos\theta). \tag{4.107}$$

We start by writing

$$\psi_{nlm}^{(2)}(\vec{r}) = \sum_k \frac{X_{nlmk}^{(2)}}{r} Y_k^m(\Omega). \tag{4.108}$$

The second-order density,

$$n^{(2)}(\vec{r}) = \sum_{nlm} \left\{2 \mid \psi_{nlm}^{(1)} \mid^2 + 4\Re[\psi_{nlm}^{(0)*}(\vec{r})\psi_{nlm}^{(2)}(\vec{r})]\right\}, \tag{4.109}$$

may be expanded as

$$n^{(2)}(\vec{r}) = \sum_h [\bar{n}_h^{(2)}(r) + \bar{\bar{n}}_h^{(2)}(r)]P_h(\cos\theta), \tag{4.110}$$

where

$$\bar{n}_h^{(2)}(r) = \frac{2h+1}{2\pi r^2} \sum_{nljj'} B^M(l, jj' \mid h) X_{nlj}^{(1)} X_{nlj'}^{(1)}, \tag{4.111}$$

$$\bar{\bar{n}}_h^{(2)} = \frac{2h+1}{\pi r^2} \sum_{nlmk} C(h0; lm \mid km) X_{nl}^{(0)}(r) X_{nlmk}^{(2)}(r). \tag{4.112}$$

The coefficient $C(h0; lm \mid km)$ has been defined in (3.55), and $X^{(n)}$ are the radial functions that vanish at the origin for all angular momentum components. The coefficient $B^M(l, jj' \mid h)$ involves the product of four 3-j symbols and a 6-j symbol; the explicit expression may be found in Senatore and Subbaswamy (1986). Performing the orthogonal transformation

$$X_{nlmk}^{(2)} = \sum_h C(h0; lm \mid km) \bar{X}_{nlhk}^{(2)}, \tag{4.113}$$

one can show that the functions $\bar{X}^{(2)}_{nlhk}$ obey the reduced second-order equation

$$\begin{aligned}
&\left(-\frac{d^2}{dr^2}+\frac{k(k+1)}{r^2}+V^{(0)}(r)-\lambda^{(0)}_{nl}\right)\bar{X}^{(2)}_{nlhk} \\
&= \delta_{h,M}C(M0,lm\mid lm)\lambda^{(1)}_{nlM}X^{(1)}_{nlk} \\
&\quad -\sum_j D^M(j,lk\mid h)V^{(2)}_M X^{(1)}_{nlj}+(\delta_{l,k}\lambda^{(2)}_{nlh}-V^{(2)}_h)X^{(0)}_{nl}. \qquad (4.114)
\end{aligned}$$

Here

$$D^M(j,lk\mid h)=\frac{(2h+1)B^M(j,lk|h)}{\sqrt{2k+1}\sqrt{2l+1}\,C(h0,l0|k0)} \qquad (4.115)$$

and

$$\begin{aligned}
\lambda^{(2)}_{nlh} &= \int_0^\infty dr\,X^{(0)^2}_{nl}v^{(2)}_h \\
&\quad +\int_0^\infty dr\,X^{(0)}_{nl}\sum_j D^M(j,ll|h)X^{(1)}_{nlj}. \qquad (4.116)
\end{aligned}$$

The first-order potential has been given explicitly in (4.10), and the second-order potential is given by

$$\begin{aligned}
V^{(2)}_h(r) &= \frac{8\pi}{2h+1}\int dr' r'^2 n^{(2)}_h \frac{r^h_<}{r^{h+1}_>} \\
&\quad +v'_{xc}(n^{(0)}(r))n^{(2)}_h(r) \\
&\quad +\frac{1}{6}v_{xc}''(n^{(0)}(r))[n^{(1)}(r)]^2\frac{2h+1}{2M+1}C(h0,M0|M0). \qquad (4.117)
\end{aligned}$$

From the properties of the 3-j and 6-j symbols (see, e.g., Rotenberg *et al.* (1959)) the nonvanishing values of h and k for given l and M can be determined. It is easily shown that h takes on the values 0,2, ..., $2M$, and for given h, k takes on the values $l+h, l+h-2, \ldots, \mid l-h \mid$.

4.4.2 Negative Ions

As already mentioned, negative ions provide a particularly stringent test of atomic theories. As discussed in Section 4.1.2, in the local density theory the discussion of negative ions requires the use of self-interaction correction, at least in the ground state. The results for some negative ions from

Table 4.15: Calculated hyperpolarizabilities of negative ions.

	$\gamma/6$ (10^{-39} esu)		
	H^-	F^-	Cl^-
PSIC[a]	0.288×10^6	0.120×10^4	0.723×10^4
HF[b]		$>0.084\times10^4$	

[a]G. Senatore and K.R. Subbaswamy, Phys. Rev. A **34**, 3619 (1986)
[b]P.W. Fowler and P.A. Madden, Phys. Rev. B **30**, 6131 (1984)

the PSIC calculations of Senatore and Subbaswamy (1986) are compared in Table 4.15 with the CHF-MP calculations by Fowler and Madden (1985). Unfortunately, there are no reliable experimental values for comparison. Previous experience indicates, however, that the HF results are too low, and the PSIC results too large; the actual values are likely to be near the average of the two.

4.4.3 Dipole-Quadrupole Susceptibility

So far we have confined our discussion to the nonlinear contributions to the dipole moment in the presence of a uniform electric field. In the presence of a nonuniform field there are contributions from the field gradients to the dipole moment (see, e.g., Buckingham, 1967). To lowest order in the field gradient one has

$$p_\mu^{NL} = \frac{1}{6}\gamma_{\mu\nu\alpha\beta}\mathcal{E}_\nu\mathcal{E}_\alpha\mathcal{E}_\beta + \frac{1}{6}B_{\mu\nu\alpha\beta}(\partial_\nu\mathcal{E}_\alpha)\mathcal{E}_\beta + \cdots. \tag{4.118}$$

Similarly, the induced quadrupole moment has the nonlinear contribution

$$q_{\mu\nu}^{NL} = \frac{1}{6}B_{\mu\nu\alpha\beta}\mathcal{E}_\alpha\mathcal{E}_\beta + \cdots. \tag{4.119}$$

The tensor $B_{\mu\nu\alpha\beta}$ is called the dipole-quadrupole tensor. Although there are no direct measurements of this quantity, it plays an important role in many nonlinear processes in the condensed phase such as Raman scattering. This will be discussed in Chapter 5. Here, we point out that this quantity is calculated very easily, as a by-product of the calculation of γ described above. First, note that for a spherically symmetric atom or ion, there is only

Table 4.16: Calculated values of static dipole-quadrupole hyperpolarizability B, in units of 10^{-40} esu.

	10^{-40} esu	
	TDLDA[a]	Other
He	-3.03	
Ne	-6.95	
Ar	-81.2	-76.6[b]
Kr	-176	
Xe	-417	
H^-	-0.242×10^5	
F^-	-318	<-198[c]
Cl^-	-1680	

[a]G. Senatore and K.R. Subbaswamy, Phys. Rev. A **34**, 3619 (1986)

[b]I. Cernusak *et al.* Chem. Phys. Lett. **128**, 18 (1986) The calculation was CHF with fourth-order many-body perturbation corrections

[c]P.W. Fowler and P.A. Madden, Phys. Rev. B **30**, 6131 (1984). The result was for CHF with a restricted basis set (no *f*-functions)

one independent component of the dipole-quadrupole tensor:

$$B_{zzzz} = B, \tag{4.120}$$

$$B_{xxzz} = -\frac{2}{3}B_{xzxz} = -\frac{1}{2}B. \tag{4.121}$$

From the expression for the quadrupole moment, (4.119), it is readily seen that

$$\frac{B}{a_0^6/e} = -2\int d^3r\, n^{(2)}(\vec{r})r^2P_2(\cos\theta). \tag{4.122}$$

We note, in passing, that $e/a_0^6 = 0.4571\times10^{-40}$ esu. Since the second-order density is calculated as a part of the calculation of γ, the determination of B involves only one additional quadrature. Senatore and Subbaswamy (1986) determined B for the rare gases and some negative ions. The calculations were performed with SIC in the unperturbed state (i.e., PSIC scheme).

Table 4.17: Low frequency dispersion in $\chi^{(3)}$. The value at λ = 10550 Å is shown relative to the static value.

	$\chi^{(3)}(3\omega)/\chi^{(3)}(0)$	
	TDLDA[a]	Expt.[b]
He	1.08	1.05
Ne	1.10	1.12
Ar	1.20	1.19
Kr	1.27	1.34
Xe	1.36	1.71

[a]Computed from the results of G. Senatore and K.R. Subbaswamy, Phys. Rev. A **35**, 2440 (1987)
[b]Computed from the results of H.T. Lehmeier *et al.*, Opt. Commun. **56**, 67 (1985)

These are tabulated, along with other calculated values, in Table 4.16. As in all such calculations, the local density results are larger than the Hartree-Fock based results.

4.4.4 Frequency Dependence

Zangwill (1983) and Senatore and Subbaswamy (1987) considered the frequency-dependent nonlinear response of closed-shell atoms in local density theory. The Green's function expressions for the frequency-dependent second-order response was already given in Section 3.4. The third-order response function is constructed in an entirely similar manner. Senatore and Subbaswamy (1987) calculated the term responsible for third harmonic generation (THG), namely, $\chi^{(3)}(\vec{r}_1, \vec{r}_2, \vec{r}_3; \omega, \omega, \omega)$. The details of the angular momentum decomposition may be found in their paper. The calculation is expected to be valid for frequencies such that both ω and 3ω are well below the single-particle ionization threshold. The ratio of their calculated $\chi^{(3)}(3\omega)$ to the static value is compared with experiment for λ = 10550 Å in Table 4.17. Our discussion in Section 2.5 indicates that TDLDA should be good for frequencies low compared to electronic resonances. This is borne out by the results shown in Table 4.17. As one goes to the heavier atoms, the first electronic resonance gets nearer to the photon frequency, and the agreement gets worse. In summary, TDLDA provides results for linear and nonlinear polarizabilities that are comparable to coupled Hartree-Fock results with many-body corrections, at substantially lower computational effort. TDLDA, even with presently used self-interaction schemes, overestimates the polarizabilities, while the Hartree-Fock based calculations tend to underestimate them. The experimental values, where known, tend towards the average of these two sets of calculations. This fact could be used to estimate the polarizabilities when experimental values are unavailable.

Chapter 5

Ionic Solids

5.1 Spherical Solid Model

In crystals the atoms are arranged in a periodic array. A single unit cell in real space is a polyhedron. Electron states in the crystal obey Bloch's theorem, which states that the wave functions can be written in the form $\psi_{j\vec{k}}(\vec{r}) = u_{j\vec{k}}(\vec{r}) \exp(i\vec{k}\cdot\vec{r})$, where the first factor is periodic in the lattice: $u_{j\vec{k}}(\vec{r}+\vec{a}_\ell) = u_{j\vec{k}}(\vec{r})$. Wigner and Seitz (1933) were interested in calculating the eigenfunction for metallic sodium for the state with $\vec{k} = 0$. For this state they noted that periodicity is satisfied by assuming that the normal derivative of the function vanishes at each surface of the polyhedron which defines the unit cell,

$$0 = \hat{n} \cdot \vec{\nabla} u_{j\vec{0}}(\vec{r}). \tag{5.1}$$

The unit vector $\hat{n}$ is perpendicular to the faces of the polyhedron. The above boundary condition does not apply when $\vec{k} \neq 0$. Periodicity does not require that the normal derivative vanish. However, it is valid for states at $\vec{k} = 0$ in solids with inversion symmetry. The solution to the band structure problem is still difficult. One must solve Schrödinger's equation for the electron with the boundary condition that the cell-periodic part of the wave function have a normal derivative which vanishes over the surface of the polyhedron.

The complexity of the band structure problem requires a computer in order to find the solution. This has caused many investigators to seek approximate methods by which an accurate answer can be obtained analytically, or with less numerical effort. One of these approximations is the *spherical solid model.*

5.1.1 Metals

Wigner and Seitz (1933) invented the method of finding electron states by solving the equation inside a polyhedron, with the boundary condition that the wave function have a normal derivative which vanishes at the surface. They also pointed out that the problem becomes very simple if the polyhedron is replaced by a sphere. They invented the spherical solid model. Indeed, for simple metals such as the alkalis the method works quite well. It gives many energy levels with an accuracy of 1%.

The volume of the sphere is set equal to the volume of the polyhedron. The radius of the sphere is $r_{WS} = Z_v^{1/3} r_s$, where the density parameter r_s was introduced in Section 3.1. Here Z_v is the valence of the metal ion, which is $Z_v = 1$ for the alkalis, $Z_v = 2$ for the alkaline earths, etc. Equation (5.1) is replaced by the condition that $\partial u / \partial r = 0$ at the surface of the sphere. In spherical coordinates the wave function of the electron can be written as

$$u_{\vec{k}}(\vec{r}) = \sum_{\ell m} C_{\ell m} R_\ell(kr) Y_{\ell m}(\theta, \phi), \tag{5.2}$$

where $C_{\ell m}$ is a constant. The angular dependence is given by the spherical harmonic functions $Y_{\ell m}(\theta, \phi)$. The unknown part is the radial function $R_\ell(kr)$. It is usually found by a numerical solution to the radial part of Schrödinger's equation. However, the spherical solid model has changed the determination of the band structure from a complicated three-dimensional calculation to a one-dimensional solution for the radial function. This latter problem is rather simple to calculate.

Many calculations have been done for metals utilizing the spherical solid model. It is impossible to review or even mention all of the published applications of this useful model. Here we shall mention one case, namely the study of electron screening around fixed charges in metals. The fixed charges could be the ion cores, impurities, or core holes in metals. The simplest physical model is to use a Yukawa potential for the screened interaction,

$$V_s(r) = \frac{e^2}{r} e^{-k_s r}. \tag{5.3}$$

Although this approximation is common, it is inaccurate. The screening charge is usually nonlinear. When the fixed charge is on an ion, the screening charge is mostly *outside* of the ion. The core hole is not a point charge but has a finite extent due to the size of the atomic orbital. The actual screened interaction is quite different from a Yukawa potential. The spherical solid model can be used with LDA to address this problem.

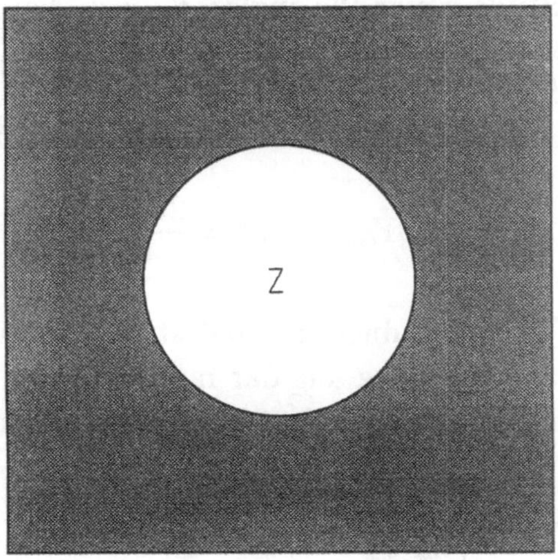

Figure 5.1: A spherical hole in jellium. The positive charge is in the shaded region. The nucleus at the origin has a point charge Z.

One model which has been well studied is illustrated in Figure 5.1. The jellium model of a metal spreads the positive charge uniformly through the solid. The positive and negative charge have equal particle densities n_0. The positive charge is fixed in place. The Wigner-Seitz cell is a spherical hole in the positive background. At the center of the hole is a point charge Z which represents the nucleus. One then solves the Kohn-Sham equations for both the bound and continuum states of the ion. All states are filled up to the chemical potential $\mu = \hbar^2(3\pi^2 n_0)^{2/3}/2m$ which then maintains the charge neutrality. The bound shells of the ion are also filled with electrons, which then provides a model of an ion in a metallic environment.

The usual practice is *not* to apply Wigner-Seitz boundary conditions that the derivative of the radial eigenfunctions vanish at the surface of the sphere. That constraint is only valid for the state at $\vec{k} = 0$. It does not apply to electron states with wave vector $\vec{k} \neq 0$. Instead, the usual practice is to assume the states with $\vec{k} \neq 0$ are standing waves. The electron density is

$$n(\vec{r}) = \sum_{j\ell(occ.)} |\psi_{j\ell}(\vec{r})|^2 + \frac{2}{V} \sum_{\vec{k}<k_F} |\psi_{\vec{k}}(\vec{r})|^2. \tag{5.4}$$

The first term is from the bound states, while the second term is from the states in the conduction band. The vector $\vec{k}$ is represented in spherical coordinates by its magnitude k plus the angles (θ_k, ϕ_k). The angular dependence is given by the spherical harmonic functions $Y_{\ell m}(\theta_k, \phi_k)$. The quantum numbers are (k, ℓ, m). Since the energy depends only upon the magnitude of k in this spherical model, the summation over m can be done immediately,

$$\sum_m |Y_{\ell m}(\theta, \phi)|^2 = \frac{2\ell+1}{4\pi}. \tag{5.5}$$

The summation over the conduction band states is now just a summation over the wave vector k plus the angular momentum ℓ. For ions with only filled electron shells, the density is just a function of the radial distance r,

$$n(r) = \sum_\ell (2\ell+1)\left[\frac{1}{2\pi}\sum_j R_{j\ell}(r)^2 + \frac{1}{\pi^2}\int_0^{k_F} k^2 dk R_\ell^2(kr)\right]. \tag{5.6}$$

The first sum is over the bound states in the ion core. The integral is over the conduction band states. For bound states, the summation over angular momentum ℓ only has a few terms, since there are only a few bound states. For the conduction band states, the summation over angular momentum must be carried to infinity in order to fill up all of the electron states at large distances from the Wigner-Seitz sphere. If the nuclear potential were missing, the solution to the Kohn-Sham equation would be just the spherical Bessel's functions $j_\ell(kr)$. One can utilize the sum rule for Bessel's functions,

$$1 = \sum_{\ell=0}^{\infty}(2\ell+1)j_\ell(kr)^2, \tag{5.7}$$

to write the density in the form

$$n(r) = n_0 + n_b(r) + \delta n_c(r), \tag{5.8}$$

$$n_b(r) = \sum_{\ell j}\frac{(2\ell+1)}{2\pi}R_{j\ell}(r)^2, \tag{5.9}$$

$$\delta n_c(r) = \frac{1}{\pi^2}\sum_\ell (2\ell+1)\int_0^{k_F} k^2 dk (R_\ell^2(kr) - j_\ell^2(kr)). \tag{5.10}$$

The density n_0 has been added as the first term, which is exactly canceled by the Bessel function term at the end. This way of writing is convenient since the electron density at large distances is n_0. The other terms give a contribution which is localized near the origin. The density from the bound states is $n_b(r)$. The excess conduction band density $\delta n_c(r)$ gives the difference from the uniform density. Now only a few terms are needed in the angular momentum summation for $\delta n_c(r)$. The states $R_\ell(kr)$ rapidly become identical to $j_\ell(kr)$ at larger values of ℓ . The difference $[R_\ell^2(kr) - j_\ell^2(kr)]$ vanishes as ℓ becomes large.

The ground state is found by solving the Kohn-Sham equations both for the bound and for the conduction band states. The states with $k_F > k > 0$ are still part of the ground state of the metal, so the LDA method should be a valid method of calculation.

The major difficulty with the calculation is the Friedel oscillations. The electron density oscillates at large distances. Far away from the impurity the radial functions have the asymptotic behavior:

$$\lim_{z\to\infty} j_\ell(z) \rightarrow \frac{\sin(z - \frac{\ell\pi}{2})}{z}, \tag{5.11}$$

$$\lim_{z\to\infty} R_\ell(z) \rightarrow \frac{\sin(z - \frac{\ell\pi}{2} + \delta_\ell)}{z}. \tag{5.12}$$

The part of the density due to core states decays to zero exponentially. The variation in density is from the oscillations in the density of the conduction electrons,

$$\delta n_c(r) \rightarrow \frac{1}{2\pi^2 r^3} \sum_\ell (-1)^\ell (2\ell + 1) \int_0^{k_F r} dz[\cos(2z) - \cos(2z + 2\delta_\ell(z))],$$

where δ_ℓ is a phase shift .

One application of this model has been to study nonlinear screening. Zaremba *et al.* (1977) set $Z = 1$ and studied the nonlinear screening charge around a proton embedded in jellium. Almbladh and von Barth (1976) used atomic values for Z, and studied the x-ray absorption in the metal. Bryant (1978, 1979) solved the problem of an electron missing from one of the inner cores of the ion. This creates a core hole. The screening charge around it is interesting. Figure 5.2 shows his numerical results for the $3p$ core hole in potassium. The quantity shown in the figure is $4\pi r^2 \delta n_c(r)$. The oscillations at small values of r are due to the ion core states. The variations at very large values of r are the start of the Friedel oscillations. One can see that the screening charge is mostly outside of the ion core. The screening charge has the same shape regardless of which state has the core hole.

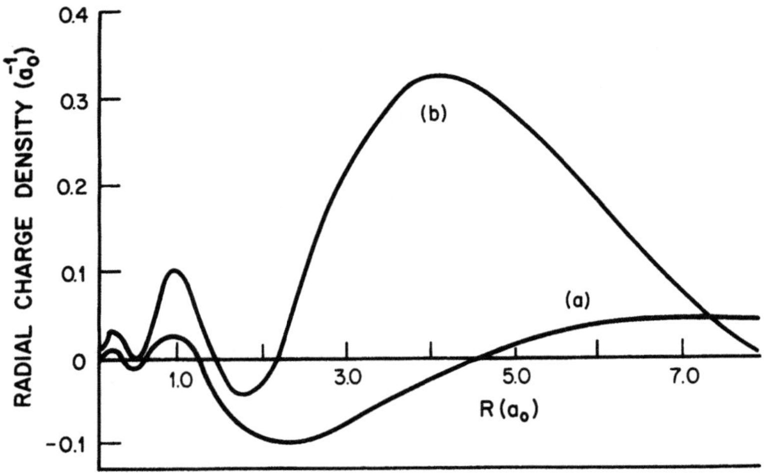

Figure 5.2: Radial charge density of the excess valence electrons in K: (a) for the ground state, and (b) for the state with a $3p$ core hole. From Bryant (1979).

5.1.2 Ionic Solids

The spherical solid model has been applied to ionic solids. The properties of a single ion are calculated by assuming that its local environment is spherical. Naturally, this produces ions with spherical charge distributions. Twenty years ago everyone thought this was a terrible model. At that time the common model for ionic crystals assumed significant bonding. Bonding distorts the charge distributions in the regions where neighboring ions are in contact (Phillips, 1973).

A calculation by Jennison and Kunz (1976) first suggested the bonding model was incorrect. They did a Hartree-Fock calculation of the band structure of several ionic solids. After they found the self-consistent charge densities in Hartree-Fock, they calculated the potential energy acting at each nuclear site in the crystal. They found that the result equaled the Madelung potential to an accuracy of 0.1%. The Madelung potential is obtained by adding up the Coulomb potential from a lattice of point charges. The same result is obtained if the charge distributions are not points, but are spherical. So Jennison and Kunz found that the charge distributions in the ionic solids are very well described by a summation of spherical distributions. The charge distributions overlap in the region between two neighboring ions. However, in this region the electron density is accurately given by adding the spherical distributions of the two neighbors.

Inspired by the result of Jennison and Kunz, Mahan (1980a) invented the spherical solid model for ionic crystals. He later used TDLDA to calculate the polarizability of ions in crystal, and was able to predict the refractive indices accurately. This was the first easy method of calculating this quantity. He showed that the halide ions are compressed in the solid, and by a different amount in different crystals.

Colella's group (1982, 1985) remeasured the charge densities in several alkali halide solids by x-ray scattering. The model which best fits their data has spherical charge distributions for each ion. They also concluded the halide ions are compressed. This measurement confirmed the predictions of Jennison and Kunz that the distributions are spherical, and of Mahan that the halide ions are compressed.

The key assumption in the spherical solid model, in ionic crystals, is that the electrons are localized on a single ion. One can solve the Kohn-Sham equations for electron states on a single ion. The electrons on the other ions contribute to a crystalline potential. This assumption is certainly an approximation. A theorem by Moyer (1974) proves there are no localized states in a system with a periodic potential. Moyer proved his theorem for a one-dimensional solid in the Hartree-Fock approximation. Neither correlation nor dimensionality is likely to change the result. The states are not localized, and assuming they are is an approximation. The error in this approximation is difficult to estimate.

An entirely different question is whether a hole in the valence band is localized. Often holes become self-trapped by strong polarization forces due to phonons and other electrons. However, in a filled valence band, the electrons are not localized.

The Kohn-Sham equations are solved for electrons in a central ion. The other ions in the solid contribute to the crystalline potential. This potential can be written as a summation over the ions in the crystal

$$V_C(\vec{r}) = \sum_{j \neq 0} V_j(\vec{r} - \vec{R}_j). \tag{5.13}$$

The spherical solid model has been applied to cubic solids. Each ion sits at a site of cubic symmetry. The crystalline potential can be expanded in Kubic Harmonics (von der Lage and Bethe, 1947). They are angular functions $\alpha_\ell(\theta, \phi)$ which characterize the site symmetry. For cubic crystals one has

$\alpha_0 = 1$ and $\alpha_\ell = 0$ for $\ell = 1, 2, 3$. The potential can be written as

$$\begin{aligned} V_C(\vec{r}) &= V_C^{(0)}(r) + V_C', \qquad (5.14)\\ V_C'(\vec{r}) &= \sum_{\ell \geq 4} \alpha_\ell(\theta, \phi) V_C^{(\ell)}(r), \qquad (5.15)\\ V_C^{(0)}(r) &= \int \frac{d\Omega}{4\pi} V_C(\vec{r}). \qquad (5.16) \end{aligned}$$

The potential $V_C^{(0)}(r)$ is spherically symmetric. It is added to the other potentials when solving for the Kohn-Sham eigenfunctions. The equation which is solved is

$$\left[-\nabla^2 - \frac{2Z}{r} + V_H(r) + V_{xc}(r) + V_C^{(0)}(r) - \lambda_j \right] \psi_j(\vec{r}) = 0. \qquad (5.17)$$

The potential terms are from: the nucleus, the Hartree potential, the exchange-correlation potential, and the spherically averaged crystalline potential. These equations are solved using the same numerical technique described in Chapter 4 for solving atomic problems. Since the crystalline potential $V_C^{(0)}(r)$ is also spherically symmetric, only the radial equation must be solved numerically. The contribution of the additional crystalline potential $V_C'(\vec{r})$ may be evaluated using perturbation theory.

The basic idea in the spherical solid model is to divide the crystalline potential into two terms. The spherically averaged part is the largest term. It is incorporated into the solution of the Kohn-Sham eigenfunctions for the ground state. These crystalline eigenfunctions are used to evaluate the remaining terms by perturbation theory. The procedure works well since the remaining terms make a small contribution to the ground state energy. The terms V_C' are interesting because they depend upon crystal structure. Their contribution to the ground state energy gives a different amount for different structures. These terms are important for determining the ionic arrangements.

How is the crystalline potential $V_C^{(0)}(r)$ calculated? One method is to represent the electron-ion interaction V_i by a pseudopotential. Usually we use a Heine-Abarenkov (1964,1965) pseudopotential which has the form:

$$V_i(r) = \begin{cases} -2Z_v/r, & r > r_i, \\ A_i, & r < r_i, \end{cases}$$

where Z_v is the valence charge of the ion. The core energy A_i actually depends upon the energy of the electron, as well as its angular momentum ℓ.

However, this dependence is usually ignored, and a value used for A_i which is averaged over the energy of s-waves. For ionic crystals, it makes sense to choose the core radius r_i to be the radius of the ion. Pauling's (1960) values are used. Then the core energy A_i is positive, and the pseudopotential is repulsive.

For positive ions the value of A_i is found by solving for the eigenvalues of electrons bound to the s-states which are outside of the closed shell. One chooses the A_i which gives the same bound state energy. Negative ions such as H^- or Cl^- have no bound states for electrons outside of the closed shell, and A_i can not be found this way. Here we usually guess a value such as $A = 1$ Ry. This value is not very important since the negative ions repel electrons. A better way would be to calculate the phase shift of an unbound electron scattering from the negative ion. The energy dependence of the phase shift could be used to construct a pseudopotential. Since the negative ions repel electrons, an alternate form, $V_i(r) = A - 2Z_v/r, r < r_i$, has also been used (Mahan, 1980a).

Figure 5.3 shows the crystalline potential $V_C^{(0)}(r)$ at the site of chlorine and potassium ions in KCl. For the negative ion the crystalline potential builds a potential box around the ion. This potential box confines the negative ion. It increases the binding energy of the electrons in the ion. Recall that negative ions are not stable in LDA, although they are in nature. This problem is eliminated in the ionic crystal since the crystalline potential creates a box which increases the binding of the electrons. In fact, the Madelung potential alone is sufficient to stabilize the negative ion. All of the electrons are bound, even in LDA, and there is no problem in finding a self-consistent solution. The shape of this box varies for different crystals. The box causes the negative ions to compress, and by different amounts in different crystals. At the origin the potential is $V_C^{(0)}(0) = -e^2\alpha/a$ where a is the lattice constant, and α is the Madelung constant. The positive ions are much less influenced by the crystal potential. Their electrons have much larger binding energies.

Another method to construct the crystalline potential is by spherical averaging of the potential from the neighbors done completely in LDA. This has the appeal of preserving the first-principles nature of the whole calculation. Further, there are systems for which suitable pseudopotentials are not available. This completely self-consistent method was attempted by Senatore and Subbaswamy (1986) in the case of rare gas crystals, and by Johnson *et al.* (1987) in the case of alkali halide crystals.

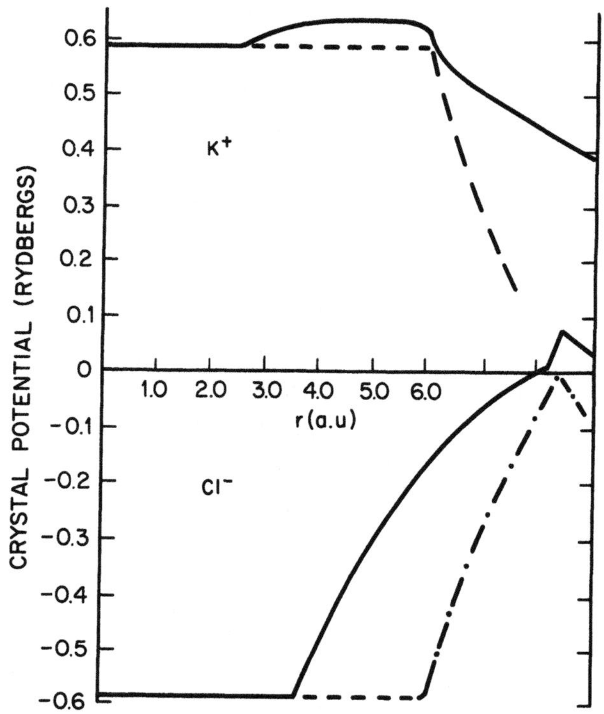

Figure 5.3: The spherically averaged crystalline potential in KCl at the K^+ and Cl^- ion sites. Dashed line is the Madelung potential obtained by representing the other ions by point charges. The solid line is from the Heine-Abarankov pseudopotential. From Mahan (1980a).

Senatore and Subbaswamy (1986) used LDA to construct a self-consistent potential for the central atom in a solid or liquid consisting of closed-shell atoms or ions. Because of the closed-shell configurations, the total electron density is split into that of the *central* atom and the remainder:

$$n(\vec{r}) = n_0(r) + \sum_{j\neq 0} n_0(\vec{r} - \vec{R}_j). \tag{5.18}$$

Such a decomposition could be formally justified in the case of a crystal in terms of Wannier type of orbitals (Senatore *et al.*, 1988). The total kinetic energy functional in (2.5) is rewritten as

$$T_s[n] = T_s[n_0] + (T_s[n] - T_s[n_0]) \equiv T_s[n_0] + U_K[n, n_0], \tag{5.19}$$

again splitting it into that of the electrons on the central atom, and the remainder. The Kohn-Sham equations are obtained by minimizing the ground state energy functional with respect to $n_0(\vec{r})$, keeping the rest of the electronic density frozen. This is done by introducing localized (atomic-like) Kohn-Sham orbitals via

$$n_0(\vec{r}) = \sum_j | \psi_j(\vec{r}) |^2 . \tag{5.20}$$

There is no formal justification for using localized orbitals, but only a preference for simplicity, in the spirit of the spherical solid model. The minimization results in the Kohn-Sham equations with an effective potential:

$$[-\nabla^2 + V(r)]\psi_j(r) = \lambda_j \psi_j(r), \tag{5.21}$$

$$V(r) = U(r) + V_H + V_{xc} + V_K, \tag{5.22}$$

$$U(r) = -\sum_j \frac{2Z}{|\vec{r} - \vec{R}_j|}. \tag{5.23}$$

The first three terms are, respectively, the nuclear potential, the Hartree potential, and the exchange-correlation potential. The last term arises from the kinetic energy of the electrons on the other atoms. The positions of the other atoms are denoted as $\vec{R}_j$. The requirement that all atoms are equivalent (in the case of a monatomic crystal) leads to the additional self-consistency requirement,

$$n_j(\vec{r} - \vec{R}_j) \equiv n_0(\vec{r}). \tag{5.24}$$

The spherical solid model is used to average the crystalline density. The density as a function of the vector $\vec{r}$ is replaced by a function of the scalar r,

$$n(r) = n_0(r) + \frac{\rho}{2r} \int d^3R \frac{g(R)}{R} \int_{|R-r|}^{R+r} r' dr' n_0(r'). \tag{5.25}$$

Here ρ is the average number density of atoms in the system, and $g(R)$ is their radial distribution function. Exactly the same type of averaging is done for the potential terms. Their value $V(n(r))$ is calculated for the central atom using the density $n(r)$. Then the same spherical potential is assumed to be on all of the other atoms, and a spherical average is done using the above formula with n_0 replaced by $V(n(r))$.

The full density $n(r)$ of the solid is used in calculating the potentials in (5.22). This choice then includes the exchange and correlation potentials of the neighboring atoms. In the spirit of LDA, Senatore and Subbaswamy (1986) also made a local *ansatz* for the potential V_K on the electrons of the central atom due to the kinetic energy of the electrons in the surrounding atoms. It is the functional derivative of the potential term in (5.19) with respect to the equilibrium density,

$$\frac{\delta U_K[n, n_0]}{\delta n_0} \equiv V_K(r) = \frac{\hbar^2}{2m}(3\pi^2)^{2/3}[n(r)^{2/3} - n_0(r)^{2/3}]. \tag{5.26}$$

Here we give the dimensional form of the potential, in order to show that this term is just the Thomas-Fermi kinetic energy discussed in Chapter 1. This procedure accounts for the fact that the effective pseudopotential on an electron outside an atom is somewhat repulsive. From the work of Gordon and Kim (1972) it is known that in LDA the repulsive potential originates from the kinetic energy term in the ground state energy.

The above scheme is entirely self-consistent. One starts with a first guess at the potential— usually the atomic potential — which is sufficient tofind $n_0(r)$ and hence $n_j(r)$. This latter density permits one to calculate a better guess at the potential $V(r)$. One uses $V(r)$ to find the new eigenfunctions and density. This process is iterated to convergence. Senatore and Subbaswamy (1986) applied this procedure to study various properties of rare gas crystals; their results will be discussed in Chapter 6.

Johnson *et al.* (1987) applied the above procedure to alkali halide crystals. The only additional complication here is that the Coulomb interaction is long-range, and hence the potential due to distant neighbors must be included. Thus, they included the electron density overlap effects only for a few shells of neighbors, while including the Coulomb potential of distant neighbors by treating them as point charges and using the Ewald procedure as before. In Figure 5.4 we compare the potential felt by an electron near a Cl^- ion in KCl calculated by this procedure with that calculated by using Mahan's pseudopotential procedure. The peak (cusp) in the self-consistent potential at the location of the nearest K^+ ion represents the overlap repulsion. Its large size can be traced to the form of the LDA approximation (5.26) for the excess kinetic energy. The large repulsion exaggerates the degree of confinement felt by the anion. Johnson *et al.* found that this results in a substantial suppression of the linear and nonlinear polarizabilities of the anions when this method is used. They concluded that Mahan's pseudopotential procedure, while introducing some empiricism into the calculation, gives far superior results for the polarizabilities.

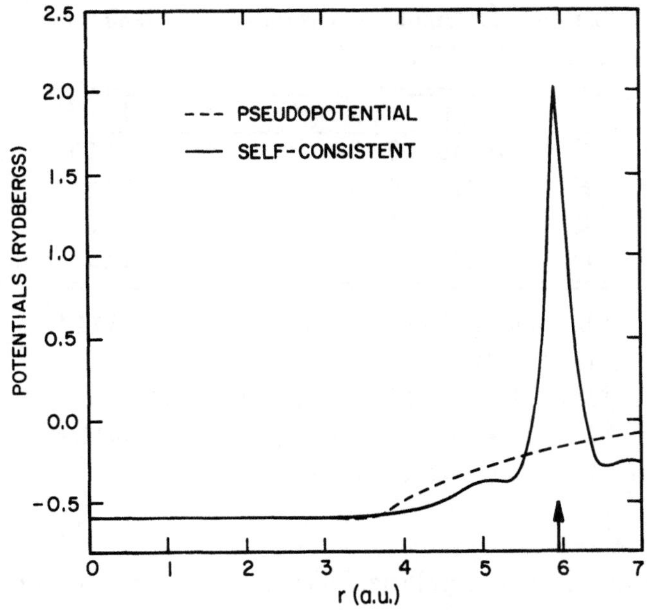

Figure 5.4: The spherically averaged self-consistent crystal potential near a Cl^- ion in KCl, compared with one calculated from pseudopotentials. From Johnson *et al.* (1987).

5.2 Dipole Polarizability

The polarizability of ions in solids is an important quantity. One reason is that they determine the refractive index n. For lattices with cubic symmetry, the relationship between α and n is called the Clausius-Mossotti formula,

$$\sum_j \alpha_{d,j} = \frac{3}{4\pi V_0} \frac{n^2 - 1}{n^2 + 2}. \tag{5.27}$$

On the left is the summation of the dipole polarizability of all of the ions in the unit cell of volume V_0. This formula is derived by including only dipole forces. It neglects electrostatic moments such as quadrupole, octupole, etc. However, they are estimated to change the answer by less than 1% (Mahan, 1980b). The polarizabilities are also useful for calculations of defect structures, phonon frequencies, surface fields, and numerous other quantities.

In ionic solids there are two low frequency dielectric constants which are denoted by ϵ_0 and ϵ_∞. At zero frequency the dielectric constant is ϵ_0. It

Table 5.1: Dipole polarizabilities of alkali ions in crystals. Units are Å^3.

	Expt.	Pauling[a]	Mayer[b]	TKS[c]
Na^+	0.148	0.182	0.182	0.255
K^+	0.81	0.86	0.84	1.20
Rb^+	1.35	1.42	1.42	1.80
Cs^+	2.34	2.45	2.45	3.14

[a]L. Pauling, Proc. Roy. Soc. (London) **A114**, 181 (1927)

[b]J. E. Mayer, J. Chem. Phys. **1**, 270 (1933)

[c]J.R. Tessman, A.H. Kahn, and W. Shockley, Phys. Rev. **92**, 890 (1953)

includes both the electronic polarizability, and the contribution from the ion displacements. The electronic contribution alone provides ϵ_∞, which is the dielectric constant at frequencies well above the vibrational frequencies of the ions, and well below any electronic resonance frequency. The refractive index at optical frequencies is $n = \sqrt{\epsilon_\infty}$. There have been many attempts to determine the dipole polarizabilities of ions in crystals. Experimentally one measures the sum of all of the polarizabilities in a cell, and some theory is needed to obtain the polarizability of the individual ions. It is an old topic with a large literature. Here we shall give a short history of the efforts to obtain the values for the alkali ions. Their values are now well known, since they have been the ions studied most frequently. In 1927 Pauling gave the values shown in Table 5.1. The first column gives experimental values. The next three columns are the values found by Pauling (1927), Mayer (1933), and Tessman, Kahn, and Shockley (1953) which is called TKS. Mayer listed Pauling as one of his sources, which is why his values are similar. The table shows that Pauling and Mayer found values which are quite close to the best values today. Those of TKS compare poorly with the experimental values.

Mayer's values are seldom quoted. His paper was on van der Waals forces. He needed the polarizability in order to estimate these forces. His paper contains an interesting statement that

> "...it has been assumed that the polarizabilities of the positive ions are known, and are constants of the ion. The polarizabilities of the negative ions have been obtained by subtracting

> from the extrapolated value of $(n^2-1)/(n^2+2)$ at $\nu=0$ for the crystal, the contribution due to the positive ion. This method is justified only by the relatively smaller contribution of the positive ion and its probably smaller variation with the crystal than that of the negative ion."

All of these assertions, which he regarded as probable, are now regarded as fact. He was the first to state that the polarizability of the positive ions is nearly the same in different solids, but that the polarizability of the negative ions varies. The electrons in the positive ions are bound tightly, and are relatively unperturbed by the crystalline environment. In contrast, the electrons in the outer shell of anions are bound loosely, and are influenced by the crystalline environment. The negative ions become compressed in the solid, and by differing amounts in each crystal.

The dipole polarizability has the dimensional units of volume. The value of the polarizability scales approximately with the volume of the ion. Large ions have large polarizabilities. When the anions are compressed in the solid, their polarizabilities are reduced. Mahan (1980a) showed the compression reduces the value of all of the polarizabilities: dipole, quadrupole, octupole, etc.

TKS made a different assumption. They assumed that all dipole polarizabilities were transferable: they were the same in each solid. Transferability was assumed for both positive and negative ions. Using this assumption they constructed a table of polarizabilities. Their values for the alkali ions in Table 5.1 are very inaccurate. The same is true for other ions. The assumption of transferability does not work well for anions.

The values proposed by TKS became standard. They were quoted in most textbooks, and used in most calculations. Only recently were their results challenged. Mahan (1980a) used the spherical solid model to calculate the polarizabilities of ions in crystals. He found the values for the alkali ions were transferable, but the values for the halide ions varied from crystal to crystal. The values scaled with the compression of the halide ion. He showed that Mayer's model was most reasonable.

It is easy to construct a table of experimental anion polarizabilities. One has to make two assumptions: (1) The Clausius-Mossotti relation is valid, and (2) the cation polarizabilities are transferable. Both of these assumptions have been tested by numerous calculations, and found to be valid.

Table 5.2: Experimental dipole polarizabilities in units of Å^3. Top row are values for the cations. Lower four rows are values for anions in the binary crystal with the cation listed above.

	Li^+	Na^+	K^+	Rb^+	Cs^+
Cation	0.030	0.147	0.81	1.35	2.34
F^-	0.89	1.02	1.20	1.18	1.27
Cl^-	2.94	3.14	3.36	3.45	3.55
Br^-	4.13	4.28	4.54	4.58	4.66
I^-	6.19	6.35	6.66	6.80	6.81

Then the anion polarizabilities in binary lattices are given by

$$\alpha_{d,a} = \frac{3}{4\pi V_0}\frac{n^2-1}{n^2+2} - \alpha_{d,c}. \tag{5.28}$$

Everything on the right is a known quantity. Table 5.2 shows this table for the halide ions in alkali halides. The halide polarizability increases in value as the alkali ion gets bigger. This behavior is also shown graphically in Figure 5.5. The polarizability is plotted *versus* the inverse lattice constant of the crystal. The points are different solids. The points on the line labeled "Cl^-" are LiCl, NaCl, KCl, RbCl. The line for F^- has five points since CsF has the NaCl structure. The line is only to guide the eye. The plots are roughly linear. Since the span of values of a is small, a linear plot is obtained regardless of whether one plots $\alpha_d \propto a$, or a^3, or a^{-3}. A larger cation makes a larger lattice constant and a larger unit cell. The anion has more space, and is compressed less. Its polarizability is larger.

The figure also shows the polarizabilities of chalcogenide ions in the alkaline earth chalcogenides: crystals such as MgO or BaSe. They also have the NaCl structure. A plot of this sort for the O^{--} ion was first given by TKS. The double negative ions are not stable as free ions in vacuum. They exist in the solid because the crystal potential increases the binding of the electrons in the outer shells. One expects their polarizability to depend upon the size of the unit cell, since the cell is holding the ion together. For these solids the polarizability of the negative ions is about ten times larger than the value for the cations. The values for the cations are not known with great accuracy, but this causes only a small error in the values for the anions. So far all of the polarizabilities mentioned in this section have been experimental values.

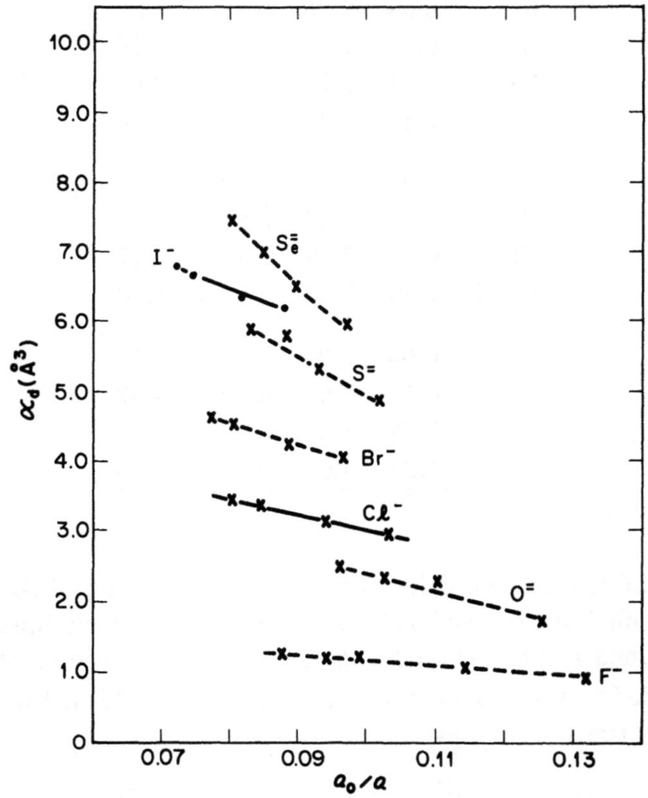

Figure 5.5: Dipole polarizabilities of anions in lattices with the NaCl structure. The units are Å^3. The horizontal axis is a_0/a, where a_0 is the Bohr radius and a is the lattice constant.

Table 5.3: Theoretical ionic polarizabilities in Å^3 calculated using TDLDA and the spherical solid model (Mahan, 1986). The symbol $\alpha_T = \alpha^+ + \alpha^-$.

	Ab initio			Expt.
Salt	α^+	α^-	α_T	α_T
LiF	0.03	0.85	0.88	0.92
NaF	0.16	1.13	1.29	1.16
KF	0.84	1.25	2.09	2.01
RbF	1.38	1.35	2.73	2.53
LiCl	0.03	2.86	2.89	2.97
NaCl	0.15	3.06	3.21	3.29
KCl	0.80	3.39	4.19	4.17
RbCl	1.30	3.52	4.82	4.80
LiBr	0.03	3.97	4.00	4.16
NaBr	0.15	3.98	4.13	4.42
KBr	0.79	4.36	5.15	5.35
RbBr	1.28	4.60	5.88	5.93

The polarizabilities can be calculated theoretically using TDLDA and the modified Sternheimer method (Mahan, 1980c). The technique is identical to that described in Sections 3.3 and 4.1, except that the crystal potential $V_C^{(0)}(r)$ is added to the ground state. For example, (3.65) is the same except that the potential $V(r)$ is now

$$V(r) = -\frac{2Z}{r} + V_H(r) + V_{xc}(r) + V_C^{(0)}(r). \tag{5.29}$$

This potential is used in calculating the Kohn-Sham eigenfunctions of the ground state, and for calculating the change in the eigenfunction due to the external field. Table 5.3 shows results obtained recently by this method (Mahan, 1986) using Gunnarsson-Lundqvist correlation. The cation polarizability does vary slightly from crystal to crystal. However, this variation is rather small compared to the variation in the anion polarizability. The variation in the polarizability of F^- is about 50% between LiF and RbF. It is interesting that the results for F^- are consistently too large, while those for Br^- are too small. The anions are strongly influenced by the crystalline potential.

The spherical solid model can be used to calculate accurately the polarizability of ions in ionic solids, and atoms in rare gas solids. The accuracy of the method is about 4-7%. This is the same error found for the calculation of the polarizability in free atoms and ions. This error is not from numerical roundoffs or uncertainties. The numerical methods produce values which are accurate to many significant figures. The error is in the TDLDA method itself. The third word in LDA is 'approximation.' Probably a better choice of correlation function would improve the results. Better results would be obtained by increasing the magnitude of the exchange-correlation potential. We resist doing this since it is *ad hoc*. A better approach would be to reexamine the formalism to identify areas of improvement. It should be noted, however, that the spherical solid approximation itself introduces an uncontrolled approximation. The calculated polarizabilities for free atoms in TDLDA are higher than experimental values, as discussed in Section 4.1. The calculated values for crystals in the spherical solid approximation are systematically lower than the experimental values. Thus, the spherical solid approximation, with the particular pseudopotentials used, appears to compensate somewhat for the rather loosely bound electrons predicted by LDA. The TDLDA results also show that Mayer's (1933) view of the polarizability is correct. The polarizabilities of the cations are transferable. The values are the same in all ionic solids. Only a small error is made using this approximation. The polarizabilities of the anions are not transferable. They vary considerably in going from crystal to crystal. Their values can be found from the Clausius-Mossotti relationship, if the values for the cations are known beforehand.

The calculation described above has been in the spherical solid approximation, where the electrons have been viewed as being localized to the ions. Of course, the electrons are actually in band states, and a correct calculation would have to use Bloch function form for the Kohn-Sham orbitals. Such a calculation was in fact performed by Hybersten and Louie (1987) for several crystals, including LiCl. They used an *ab initio* pseudopotential approach, and treated the valence states in a plane wave expansion. The dielectric matrix is calculated in a perturbative approach which is equivalent to the TDLDA formulation. The sum over unoccupied orbitals is explicitly carried out to desired accuracy. We will return to this method in Chapter 6. Hybersten and Louie (1987) found ϵ_∞ for LiCl to be 3.07, corresponding to α_T of 3.30 Å^3. This should be compared with the value 2.89 Å^3 found in the spherical solid approximation (see Table 5.4). Thus, the spherical solid approximation, with the empirical pseudopotential, gives better results at a

Table 5.4: Parameters in the Wilson-Curtis expression for anion polarizabilities in alkali halide crystals determined by fitting to TDLDA results. From Adair *et al.* (1989).

	α_0^- (Å^3)	C (Å^2)
F^-	2.24	3.87
Cl^-	5.50	4.36
Br^-	7.24	4.66
I^-	10.5	5.41

fraction of the cost of doing a full-fledged band structure calculation.

The anion polarizabilities for alkali halides obtained in TDLDA and listed in Table 5.4 can be parametrized by a simple expression due to Wilson and Curtis (1970):

$$\alpha^- = \alpha_0^- \exp(-C/R^2), \tag{5.30}$$

where α_0^- corresponds to the free ion polarizability, R is the nearest neighbor distance, and C is a constant for a particular anion. This model is based on a semiclassical argument within perturbation theory. Adair *et al.* (1989) used the calculated TDLDA polarizabilities of Johnson *et al.* (1987), which differ from those in Table 5.4 only slightly due to the use of Ceperley-Alder correlation, to test the Wilson-Curtis model. They found the fit to be excellent, and determined the parameters α_0^- and C. Their values are shown in Table 5.4. The Wilson-Curtis formula, with the parameters listed in Table 5.4 provides a useful parameterization of anion polarizabilites in alkali halide crystals.

5.3 Volume Dependence of Polarizability

As already stated, the static linear polarizabilities, α^+ and α^-, of the individual ions in an alkali halide crystal cannot be measured directly, but they can be related to the high-frequency dielectric constant, ϵ_∞, through the Clausius-Mossotti relation, (5.28), which we rewrite here as

$$\frac{\epsilon_\infty - 1}{\epsilon_\infty + 2} = \frac{4\pi}{3V_0}(\alpha^+ + \alpha^-), \tag{5.31}$$

where V_0 is the volume of the unit cell.

A simple relation for the volume dependence of the dielectric constant can be derived from equation (5.31) by assuming that the polarizabilities are independent of volume, which yields

$$\frac{R}{\epsilon_\infty}\frac{d\epsilon_\infty}{dR} = -\frac{(\epsilon_\infty + 2)(\epsilon_\infty - 1)}{\epsilon_\infty}, \tag{5.32}$$

where R is the nearest neighbor distance. It has been shown (Sharma *et al.* 1976; Aggarwal and Szigeti, 1970) that equation (5.32) does not agree very well with experimental data and the conclusion has been drawn that the change in polarizabilities with volume is significant. Equation (5.32) must then be replaced by

$$\frac{R}{\epsilon_\infty}\frac{d\epsilon_\infty}{dR} = -\frac{\epsilon_\infty + 2}{\epsilon_\infty}\left(\epsilon_\infty - 1 - \frac{2\pi(\epsilon_\infty + 2)}{9R^2}\left(\frac{d\alpha^+}{dR} + \frac{d\alpha^-}{dR}\right)\right). \tag{5.33}$$

The volume derivative of the dielectric constant is also related to the elasto-optic coefficients, p_{1n} (Nye, 1957). These are defined as

$$\Delta\left(\frac{1}{\epsilon_\infty}\right) = p_{1n}\tau_n, \tag{5.34}$$

where τ_n is the strain tensor in the standard contracted notation. For the NaCl structure the number of elasto-optic constants reduces exactly as the corresponding elastic constants, which is not true for all cubic crystal systems (Nye, 1957). In the same way as the bulk modulus is obtained from the elastic constants in cubic crystals, we obtain

$$V\frac{\partial(1/\epsilon_\infty)}{\partial V} = \frac{p_{11} + 2p_{12}}{3}. \tag{5.35}$$

From this we get the expression for the derivative of the high-frequency dielectric constant

$$\frac{R}{\epsilon_\infty}\frac{d\epsilon_\infty}{dR} = -\epsilon_\infty(p_{11} + 2p_{12}). \tag{5.36}$$

The volume dependence of the dielectric constant is important in the context of lattice dynamical calculations as well. The calculations of the phonon dispersion relations in the alkali halide crystals require, among other things, the polarizabilities of the individual ions (see the following section for more details). Knowing how the phonon energies depend on volume makes it then possible to calculate the pressure dependence of properties such as the thermal conductivity (Pettersson, 1989) and the heat capacity.

Most theoretical work (Sharma *et al.* 1976; Van Vechten, 1969; Hans *et al.* 1988) on the photoelastic effect in the alkali halides has been based on phenomenological assumptions. The only first-principles calculation we are aware of is that of Pettersson and Subbaswamy (1990), who calculated the polarizabilities for 16 alkali halides for a range of different values of the lattice constant in each crystal using the method described in the previous section. Specifically, for the cations they used Heine-Abarenkov pseudopotentials

$$V_{pseudo,c}(r) = \begin{cases} -2/r, & r > r_i \\ A_i, & r < r_i \end{cases} \tag{5.37}$$

and for the anions they used

$$V_{pseudo,a}(r) = \begin{cases} 2/r, & r > r_i \\ 2/r + A_a, & r < r_i. \end{cases} \tag{5.38}$$

They used values for A_i and r_i that are given in Mahan (1980a) and for A_a a value of 0.1 Ry following Johnson *et al.* (1988). These pseudopotentials were used for the nearest four neighboring shells and the more distant ions were represented as point charges. The Ceperley-Alder exchange-correlation potential was used.

The calculated polarizabilities for lattice constant corresponding to the volume at ambient pressure are given in Table 5.5 together with values for their volume derivatives. As has been noted before, the anions have a larger polarizability than the cations since the anions are larger and the electrons are more losely bound. The difference between $d\alpha^+/dR$ and $d\alpha^-/dR$ is even greater than the differences in polarizabilities. The electrons on the cation are so tightly bound that the polarizability does not change when the lattice constant is altered. This is consistent with the fact that the cations have approximately the same polarizability in all crystal environments. The anions, on the other hand, show a large dependence on volume. To a very good approximation changes in the polarizability of these crystals are totally due to changes in the anion polarizability. In Table 5.5 the calculated and experimental values of $(R/\epsilon_\infty)d\epsilon_\infty/dR$ are also given. Since experimental values of $(R/\epsilon_\infty)d\epsilon_\infty/dR$ are scarce, values derived from measured elasto-optic coefficients are also listed in Table 5.5. The values obtained using equation (5.36) and tabulated values of p_{11} and p_{12} from Landölt-Bornstein (1978) are given in the last column of Table 5.5. Values for the rubidium

Table 5.5: Calculated static linear polarizabilities, their strain derivatives and values for $\frac{R}{\epsilon_\infty}\frac{d\epsilon_\infty}{dR}$ for 16 alkali halide crystals. Two sets of experimental data for $\frac{R}{\epsilon_\infty}\frac{d\epsilon_\infty}{dR}$ are given, as explained in the text.

Crystal	α^+ Å^3	α^- Å^3	$\frac{d\alpha^+}{dR}$ Å^2	$\frac{d\alpha^-}{dR}$ Å^2	$\frac{R}{\epsilon_\infty}\frac{d\epsilon_\infty}{dR}$ Theory	$\frac{R}{\epsilon_\infty}\frac{d\epsilon_\infty}{dR}$ Expt.	$\frac{R}{\epsilon_\infty}\frac{d\epsilon_\infty}{dR}$ Eq. (5.36)
LiF	0.032	0.848	0.001	0.96	-0.47	-0.57[a]	-0.55
LiCl	0.032	2.81	0.001	2.05	-1.09	-	-
LiBr	0.032	3.86	0.001	2.63	-1.23	-	-
LiI	0.032	5.67	0.001	3.40	-1.53	-	-
NaF	0.158	1.13	0.006	1.07	-0.62	-	-0.84
NaCl	0.158	3.26	0.008	2.04	-1.13	-0.95[b]	-0.98
NaBr	0.158	4.40	0.006	2.57	-1.28	-	-1.35
NaI	0.159	6.37	0.006	3.29	-1.53	-	-
KF	0.839	1.28	0.031	1.08	-0.99	-	-1.22
KCl	0.838	3.50	0.029	1.98	-1.21	-0.93[b]	-1.09
KBr	0.838	4.66	0.030	2.41	-1.33	-1.42[c]	-1.29
KI	0.838	6.68	0.030	3.10	-1.49	-1.60[c]	-1.47
RbF	1.39	1.38	0.063	1.09	-1.27	-	-
RbCl	1.39	3.68	0.050	1.90	-1.38	-1.44[d]	(-1.38)
RbBr	1.39	4.89	0.051	2.38	-1.44	-	(-1.54)
RbI	1.38	6.95	0.054	3.00	-1.59	-	(-1.57)

[a] Vedam and Schmidt (1965)

[b] Burstein and Smith (1948)

[c] Vedam *et al.* (1969)

[d] Vedam and Schmidt (1966)

salts are in parentheses since the tabulated elasto-optic constants are not strict experimental values as explained in Pettersen (1981).

The calculated values for $(R/\epsilon_\infty)d\epsilon_\infty/dR$ agree well with experiment. The differences between values obtained in different experiments are almost as large as the differences between theory and experiment. There appears to be no systematic deviation between theory and experiment.

It has been shown in experiment that the refractive index varies linearly with volume strain in KBr, KI (Vedam *et al.,* 1969), and RbCl (Vedam and Schmidt, 1966). Pettersson and Subbaswamy (1990) also calculated the refractive index for RbCl as a function of volume strain all theway to the volume at which a phase transition to the CsCl structure is known to occur. The calculations give a perfect linear relationship between the change in the refractive index and the volume strain which is shown in Figure 5.6 (crosses). The straight line in Figure 5.6 is drawn with the slope obtained from experiment (Vedam and Schmidt, 1966). There is very good agreement between theory and experiment. This agreement lends support to the validity of LDA and the spherical solid model in calculations of the polarizability and its volume dependence.

It is interesting to compare the first-principles calculation with the phenomenological theory used by Sharma *et al.* (1976). They calculated the derivatives of the polarizabilities from a theory developed by Ruffa (1963), in which the cation polarizability scales with the Madelung energy according to

$$\frac{\alpha^+(R)}{\alpha_f^+} = \left(\frac{E_f}{E_f - V_M(R)}\right)^2, \tag{5.39}$$

where α_f^+ is the free-state polarizability, E_f is an empirical free-state energy parameter and $V_M(R)$ is the Madelung potential.

The derivative of the cation polarizability, $d\alpha^+/dR$, is easily obtained by taking the derivative of equation (5.39), since, on the right-hand side, it is only the Madelung potential that depends on the lattice constant. In order to obtain the corresponding derivative for the anion, Sharma *et al.* assumed that the polarizability of an ion is proportional to its radius, i.e.,

$$\frac{\Delta\alpha^+}{\alpha^+} = \frac{\Delta r_+}{r_+}, \tag{5.40}$$

and

$$\frac{\Delta\alpha^-}{\alpha^-} = \frac{\Delta r_-}{r_-}, \tag{5.41}$$

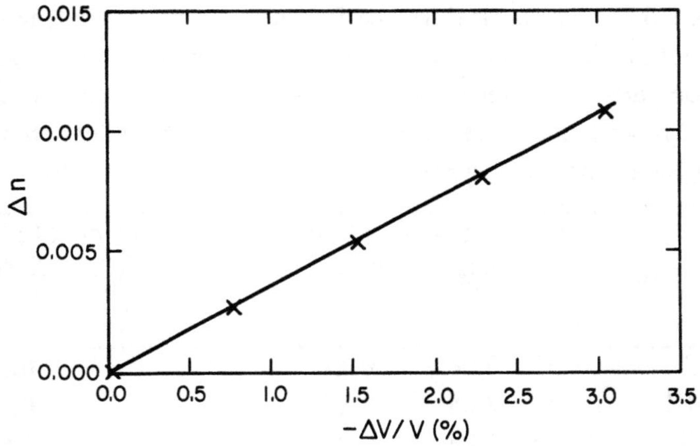

Figure 5.6: The change in the refraction index as a function of volume strain for RbCl. The crosses are theoretical values and the straight line is drawn with the slope that was obtained in experiment (Vedam and Schmidt, 1966). RbCl transform from the rocksalt structure to the CsCl structure at a volume strain of 3.05%

where r_+ and r_- are the ionic radii. These two relations together with the assumption that

$$r_+ + r_- = R \tag{5.42}$$

give

$$\frac{r_-}{\alpha^-}\frac{d\alpha^-}{dR} + \frac{r_+}{\alpha^+}\frac{d\alpha^+}{dR} = 1, \tag{5.43}$$

from which $d\alpha^-/dR$ can be obtained.

In Table 5.6 the first principles values for $d\alpha^+/dR$ and $d\alpha^-/dR$ of Pettersson and Subbaswamy are compared with the values calculated by Sharma *et al.* (1976). The values obtained for $d\alpha^-/dR$ in the two calculations are of roughly the same size. The most noticeable difference is the sign of $d\alpha^+/dR$. The magnitude of the polarizability is closely related to the size of the ion. The negative sign of $d\alpha^+/dR$ obtained by Sharma *et al.* indicates that the anion contracts while the cation expands when the crystal as a whole is compressed. On the other hand, the first-principles calculations indicate that both the cation and anion contract when the lattice constant decreases.

Table 5.6: Theoretical values for the volume dependencies of the polarizabilities and of the high-frequency dielectric constant. A comparison between values obtained from the present first-principles calculation (Pettersson *et al.*, 1990) and values obtained with the phenomenological theory of Sharma *et al.* (1976).

Crystal	Pettersson *et al.* (1990)			Sharma *et al.* (1976)		
	$\frac{d\alpha^+}{dR}$ Å^2	$\frac{d\alpha^-}{dR}$ Å^2	$\frac{R}{\epsilon_\infty}\frac{d\epsilon_\infty}{dR}$	$\frac{d\alpha^+}{dR}$ Å^2	$\frac{d\alpha^-}{dR}$ Å^2	$\frac{R}{\epsilon_\infty}\frac{d\epsilon_\infty}{dR}$
LiF	0.001	0.96	-0.47	-0.007	0.68	-0.57
NaCl	0.008	2.04	-1.13	-0.021	1.78	-0.91
KCl	0.029	1.98	-1.21	-0.153	1.98	-1.09
RbF	0.063	1.09	-1.27	-0.299	0.93	-1.74

This has been checked by calculating the root-mean-square radius of the ions.

Several authors (Sharma *et al.*, 1976; Ruffa, 1963; Krishnan *et al.*, 1974; Petrashen' *et al.*, 1961) have suggested that the cation radius increases when the free cation is placed in an alkali halide crystal. The suggested increase of the cation radius is due to the Madelung energy at the cation site which raises the electronic energy levels which would correspond to a more extended wave function. However, there is also the Pauli effect which tends to expel the wave function from the regions with high electron density at the neighboring anions. The Pauli effect was neglected in the phenomenological theories.

To test the effect of the Madelung potential alone Pettersson and Subbaswamy (1990) did a calculation for RbF with all surrounding ions represented as point charges. The change of lattice constant affects the polarizability of the Rb^+ ion only through the change in the Madelung potential. In this approximation they found that the cation does expand when the lattice constant decreases, with $d\alpha^+/dR = -0.050$ Å^2. This is the same sign as was obtained in the phenomenological theory of Sharma *et al.* (1976), but there is a factor of 6 difference in the magnitudes.

We do not believe that the cation in reality expands when the crystal as a whole is compressed. It should be remarked, however, the first-principles calculations do not unambiguously determine the relative importance of the Madelung potential versus the Pauli effect. This is due to the use of pseudopotentials with fixed radii to represent the neighbors. Thus, the calcula-

Madelung potential versus the Pauli effect. This is due to the use of pseudopotentials with fixed radii to represent the neighbors. Thus, the calculation does not take account of the compression of the neighbors while calculating the orbitals of the central ion. This is a second-order effect, and should be quite small. The anion polarizabilities are not affected to any great extent by the use of cation pseudopotentials since the cation radius changes very little when the lattice gets compressed. The relative error in the values for $d\alpha^-/dR$ and hence in $(R/\epsilon_\infty)d\epsilon_\infty/dR$, due to the use of volume-independent pseudopotentials is expected to be very small.

5.4 Direct Phonon Calculations

5.4.1 Introduction

Ions in the crystal vibrate with frequencies which are denoted as $\omega_\lambda(\vec{k})$. The wave vector is $\vec{k}$ and the mode index is λ. For binary crystals there are six modes: two transverse acoustical (TA), one longitudinal acoustical (LA), two transverse optical (TO), and one longitudinal optical (LO). The ion vibrations obey Newton's laws, since the ions are too heavy to require a description using quantum mechanics. The only possible exceptions to this statement among ionic solids are those made with hydrogen and lithium. That possibility will not be considered here. Quantum mechanics enters the vibrational problem by quantizing the number of phonons which are made at any moment in time. These quantized vibrations are called 'phonons.' Here our concern is with the theoretical calculation of the phonon frequencies.

5.4.2 Frozen Phonons

LDA has been used in several different methods of calculating the phonon frequencies. In what follows we describe the various LDA schemes used to deduce phonon frequencies, with emphasis on TDLDA methods. In the 'frozen phonon' method, the first step is to calculate the ground state energy E_G of the solid with all ions in their equilibrium position. Then the ground state energy is calculated when there is a lattice distortion of amplitude u. The difference between the two ground state energies can provide direct information about the energies of phonons at symmetry points in the Brillouin zone of the crystal.

Wendel and Martin (1979) did this 'frozen phonon' calculation for silicon. The two sublattices are displaced an amplitude u along the [111] direction.

Table 5.7: Comparison between experimental phonon frequencies in silicon with those calculated by the frozen phonon technique. Phonon frequencies ν given in THz. From Yin and Cohen (1980).

	LTO(Γ)	TA(X)	TO(X)	LOA(X)
ν_{cal}	15.16	4.45	13.48	12.16
ν_{exp}	15.53	4.49	13.90	13.32

The difference in the two ground state energies can be written as

$$E_G(u) - E_G(0) = \frac{1}{2}\left(\frac{1}{2}m_{Si}\right)\omega_{TO}^2 u^2 + k_{xyz}\left(\frac{u}{\sqrt{3}}\right)^3 + \cdots. \tag{5.44}$$

Stability requires that the term linear in u vanish. The crystal symmetry allows the third-order term with the coefficient k_{xyz}. The silicon reduced mass is $\frac{1}{2}m_{Si}$. The term quadratic in u provides a direct calculation of the frequency $\omega_{TO}(\Gamma)$ of the TO phonon at the Γ point, $\vec{k} = 0$. This distortion preserves the periodicity of the solid - with a rearranged unit cell. Thus the calculations are relatively easy. Wendel and Martin show that other static displacements provide information on $\omega_{TA}(X)$ and the difference in the elastic constants $C_{11} - C_{12}$. Similar calculations were done by Yin and Cohen (1980) whose results are shown in Table 5.7.

Another method is to calculate a complete dispersion curve for all of the phonons. This is acomplished by a procedure which involves calculating the ground state energy while displacing only one or two atoms at a time. One ion at site j is displaced a small distance $\vec{Q}_j$, and the ground state energy $E_G(\vec{Q}_j)$ is calculated for the solid with one ion displaced. This is subtracted from the ground state energy of the undistorted lattice. The difference between these two quantities gives an energy term proportional to Q_j^2. The constant of proportionality is one half of the force constant K_j for displacing that ion

$$E_G(\vec{Q}_j) - E_G(0) = \frac{1}{2}K_j Q_j^2 + O(Q^4). \tag{5.45}$$

This calculation is done for each different ion in the unit cell of the ionic solid. Terms of $O(Q^4)$ are neglected.

The next step is to displace two different ions. Then one can find the force tensor $K_{ij,\mu\nu}$ which couples the displacement of two different ions:

$$E_G(\vec{Q}_i, \vec{Q}_j) - E_G = \frac{1}{2}[K_i Q_i^2 + K_j Q_j^2 + 2K_{ij,\mu\nu}Q_{i\mu}Q_{j\nu}]. \tag{5.46}$$

This calculation is done for each pair of nearby ions. This calculation obtains a complete description of the force constants between pairs of atoms. That is the important quantity needed for the calculation of all of the phonon frequencies. The numerical computation to find the force constants is hard. It requires a lot of computing power to find the force constants this way. When one or two ions are displaced from equilibrium, the periodicity of the solid is broken. The calculation of the ground state energy becomes difficult. The virtue of this method is that all of the numerous polarization forces in the solid are included in the calculation. At least one includes those described well using LDA. Usually the short range polarization forces are difficult to describe well. However, they are quite important in the determination of the dynamical matrix and the phonon frequencies.

The frozen phonon method has an analogy in atomic physics. One could calculate the polarizability of an atom by actually putting an electric field $\mathcal{E}$ on the atom. The ground state energy of the atom is solved with and without the field. The difference between these two quantities gives the polarizability α,

$$E_G(\mathcal{E}) - E_G(0) = -\frac{1}{2}\alpha\mathcal{E}^2 + O(\mathcal{E}^4). \tag{5.47}$$

This method is also numerically complicated. The static field breaks the spherical symmetry, so the calculation of the ground state energy with the field $E_G(\mathcal{E})$ is laborious.

5.4.3 Sternheimer Method

The polarizability was calculated in the previous sections and chapters. A method was used which is different from calculating the ground state energy. Instead, the induced moment was found in response to an infinitesimal field. The same technique can be applied to the calculation of the phonon frequencies. Baroni *et al.* (1987) used the modified Sternheimer method to calculate the infinitesimal change in the ground state energy in response to an infinitesimal displacement of one sublattice. For silicon they found ω_{LTO} = 15.4 THz which is more accurate than the frozen phonon method shown in Table 5.7. They found equally good results for GaAs.

They have also extended the method to calculate piezoelectric constants (Gironcoli *et al.*, 1989). The change in free energy δE_G for a piezoelectric crystal with a strain σ and a field $\mathcal{E}$ is

$$\delta E_G = -\frac{1}{2}\{\text{-}\ \boldsymbol{\sigma}\cdot \mathbf{c}^{\mathcal{E}}\cdot\boldsymbol{\sigma} + \boldsymbol{\mathcal{E}}\cdot\boldsymbol{\alpha}^{\sigma}\cdot\boldsymbol{\mathcal{E}} + \boldsymbol{\sigma}\cdot\mathbf{e}\cdot\boldsymbol{\mathcal{E}}\}. \tag{5.48}$$

The elastic constant is $c^{\mathcal{E}}_{\alpha\beta}$, the dielectric susceptibility is $\alpha^{\sigma}_{\mu\nu}$, and the piezoelectric constant is $e_{\mu\alpha}$, where the index (α, β) runs over six possibilities (xx, yy, zz, xy, xz, yz) while (μ, ν) runs over three vector components (x, y, z). The superscripts denote that the elastic constants are taken at constant electric field, while the susceptibility is taken at constant strain. This notation is similar to the heat capacity of gases, where C_V, C_P denote the quantity at constant volume or pressure. In piezoelectric crystals, because of the strong coupling between electrical and mechanical forces, these coefficients do depend upon the boundary conditions. The stress X_α and the electric polarization P_μ are obtained by taking a functional derivative with respect to the strain, and the field, respectively:

$$X_\alpha = \frac{\delta(\delta E_G)}{\delta \sigma_\alpha} = c^{\mathcal{E}}_{\alpha\beta}\sigma_\beta - e_{\alpha\mu}\mathcal{E}_\mu, \tag{5.49}$$

$$P_\mu = -\frac{\delta(\delta E_G)}{\delta \mathcal{E}_\mu} = \alpha^{\sigma}_{\mu\nu}\mathcal{E}_\nu + e_{\alpha\mu}\sigma_\alpha. \tag{5.50}$$

There are two infinitesimal perturbations on the crystal: the strain σ and the field $\mathcal{E}$. They produce two infinitesimal responses: the stress X and the polarization P. Linear response is used to calculate the linear coefficients: elastic constant, polarizability, and piezoelectric constant. Gironcoli *et al.* (1989) use this method to calculate piezoelectric constant e_{14} for all III-V compounds to within the experimental error of the data.

This calculation can also be done in the spherical solid model using LDA. An ion is given an infinitesimal displacement. The goal is to calculate the infinitesimal forces on the other ions due to this displacement. This procedure is described in Section 5.5.

5.4.4 Unified Molecular Dynamics

Car and Parrinello (1985) presented a method that combines density functional theory (LDA) and molecular dynamics. It provides information on both static and dynamic properties of solids. They use classical mechanics to describe the motion of the ions, and employ the Born-Oppenheimer approximation to separate nuclear and electronic coordinates.

The motion of the ions is governed by the equation

$$M_I \frac{d^2 \vec{R}_I}{dt^2} = -\vec{\nabla}_{R_I} E_G, \tag{5.51}$$

where E_G is the ground state energy functional found in LDA. The ground state energy is obtained by a minimization technique using simulated annealing techniques (Kirkpatrick *et al.*, 1983).

To find a particular phonon frequency, the ions are initially displaced according to the eigenvector of the phonon normal mode. Then a molecular dynamics calculation is performed. The ions undergo oscillations, which are somewhat anharmonic, from which the frequency of the phonon is deduced. For the ω_{LTO} mode in silicon Car and Parrinello (1985) deduce a frequency of 20 THz, which is in good agreement with the frozen-phonon calculations. This method is very promising as a means of obtaining accurate microscopic information on solids and liquids.

5.5 Phonon Force Constants

5.5.1 Introduction

The *shell model* has been the dominant theory of phonons in ionic crystals. It was proposed by Dick and Overhauser (1958). The model has the virtue that ionic polarization is included in a simple way. Each ion is described as having a nucleus and a shell of electronic charge. The charge on the nucleus is an adjustable parameter — the shell charge plus the nuclear charge equals the ion valence. The nucleus and the shell are connected by a harmonic spring — the spring constant is another parameter. Finally, the shells on neighboring atoms also interact with harmonic springs. The first version of the model had the shell be a rigid sphere. Later versions of the theory, such as the 'breathing shell model,' permitted radial oscillations of the shell (see Schröder (1966) and Sangster (1967)). Phonon dispersion relations are the dependence of phonon frequency upon wave vector $\vec{k}$. They are measured by inelastic scattering of phonons or photons. Virtually every phonon dispersion relation for ionic solids has been fitted to the shell model. This work is reviewed by Cochran (1971), and by Bilz and Kress (1979).

Hardy (1959, 1962) and Karo and Hardy (1963) invented the *deformation-dipole model.* It is another theory of short-range polarization forces in ionic solids. A number of phonon calculations have also been done with this model. A good review is in the book by Hardy and Karo (1979). This model is an exact description of the mutual polarization between pairs of neighboring ions.

Mahan and Mostoller (1986) developed a model of three body forces in ionic crystals. They called it *indirect ionic interactions*, which is abbreviated I^3. It is a natural companion to the deformation-dipole (DD) model. DD is a model of two-body polarization, while I^3 is a model of three-body polarization. Together they provide an accurate model for lattice dynamics for ionic crystals. This combined model is called D^2I^3.

LDA can be used to calculate the parameters of the D^2I^3 model. This procedure is described in the following sections. The method is similar to the technique for solving for the polarizability. One calculates the polarization of an atom due to the force caused by the displacement of a neighboring atom. This model has an advantage over the shell model, in that the parameters of the model can be calculated from first principles, using LDA.

5.5.2 Dynamical Matrix

The frequencies of the phonons are calculated by using the following method (see, e.g., Maradudin *et al.*, 1971). The force constants K_{ij} between pairs of atoms or ions are assumed to be known. Let $E_G(\vec{Q}_1, \ldots, \vec{Q}_N)$ be the ground state energy when all N ions are displaced. The equations of motion from Newton's laws are

$$m_j \ddot{Q}_{j\nu} = -\frac{\partial E_G(\vec{Q}_1, \ldots, \vec{Q}_N)}{\partial Q_{j\nu}}, \tag{5.52}$$

$$m_j \omega^2 Q_{j\nu} = \delta_{\mu\nu} K_j Q_{j\nu} + \sum_{i \neq j} K_{ij,\mu\nu} Q_{i\mu}. \tag{5.53}$$

These equations are simple to solve. The phonon frequency is ω. One transforms the ion displacements to collective coordinates. There is a separate coordinate $\vec{Q}^{(\alpha)}_{\vec{k}\lambda}$ for each different ion α in the unit cell. The site label $j = (\ell, \alpha)$ where ℓ labels the unit cell located at R_ℓ. The location of the ion within the unit cell is $\vec{\tau}_\alpha$,

$$Q_{\ell,\alpha,\mu} = \frac{1}{\sqrt{N}} \sum_{\lambda,\vec{k}} Q^{(\alpha)}_{\vec{k}\lambda,\mu} e^{i\vec{k}\cdot(\vec{R}_\ell + \vec{\tau}_\alpha)}. \tag{5.54}$$

This definition of the collective coordinates is inserted into Newton's equations (5.53). The equations become simplified because a separate set of equations is found for each wave vector $\vec{k}$:

$$\omega^2 m_\alpha Q^{(\alpha)}_{\vec{k}\mu} = \delta_{\mu\nu} K_\alpha Q^{(\alpha)}_{\vec{k}\mu} + \sum_{\beta,\nu} \widetilde{K}_{\alpha\beta,\mu\nu} Q^{(\beta)}_{\vec{k}\nu}, \tag{5.55}$$

$$\widetilde{K}_{\alpha\beta,\mu\nu} = \sum_{\ell} K_{\ell,\alpha\beta,\mu\nu} e^{i\vec{k}\cdot(\vec{R}_\ell + \vec{\tau}_\beta - \vec{\tau}_\alpha)}. \tag{5.56}$$

In the above summation, the first ion has labels $i = (0, \alpha)$ while the second ion has the label $j = (\ell, \beta)$. The set of equations has dimensionality $3L$, where L is the number of ions in the unit cell. For binary solids the equations have a dimensionality of six. The numerical problem is now reduced to finding the eigenvalues ω^2 of the determinantal equation

$$0 = \det|\omega^2 \delta_{\mu\nu}\delta_{\alpha\beta} - \mathcal{D}_{\alpha\beta,\mu\nu}(\vec{k})|,$$
$$\mathcal{D}_{\alpha\beta,\mu\nu}(\vec{k}) = \frac{1}{m_\alpha}\{K_\alpha \delta_{\mu\nu}\delta_{\alpha\beta} + \widetilde{K}_{\mu\nu,\alpha\beta}(\vec{k})\}.$$

The frequencies $\omega_\lambda(\vec{k})$ of the phonons are the $3L$ values ω. The tensor $\mathcal{D}$ is called the *dynamical matrix*. The first step in finding the eigenvalues is to divide each row of the dynamical matrix by the ion mass m_α. This makes the dynamical matrix unsymmetric and non-Hermitian. This unpleasant feature is avoided by redefining the displacement as $q_\alpha = \sqrt{m_\alpha} Q_\alpha$. This step changes the dynamical matrix into one which is Hermitian,

$$0 = \det|\omega^2 \delta_{\mu\nu}\delta_{\alpha\beta} - \mathcal{D}_{\alpha\beta,\mu\nu}(\vec{k})|, \tag{5.57}$$
$$\mathcal{D}_{\alpha\beta,\mu\nu}(\vec{k}) = \frac{1}{m_\alpha} K_\alpha \delta_{\alpha\beta}\delta_{\mu\nu} + \sum_\beta \frac{1}{\sqrt{m_\alpha m_\beta}} \widetilde{K}_{\alpha\beta,\mu\nu}(\vec{k}). \tag{5.58}$$

For binary solids the six-dimensional determinant is easy to diagonalize. The hard part of the calculation is the determination of the force constants.

5.5.3 Polarizable Ion Model

The theory of lattice dynamics for ionic solids is developed here in stages. The first stage is the *polarizable ion model.* It is part of every theory of lattice dynamics. This theory is easy to understand, and is easy to construct the dynamical matrix for this case. It forms the basis of all models of lattice dynamics for ionic solids. The various models all have the polarizable ion model (PIM) plus additional terms toaccount for ionic polarization. In the PIM the ions are treated as polarizable point charges interacting via central forces. There are both direct and indirect interactions among atoms in a crystal. One direct term is obtained by treating the ions as point charges which interact via a Coulomb potential. At large separations R the interaction is $e^2 Z_1 Z_2 / R$, where $Z_{1,2}$ are the net charge of the ions. The Coulomb potential contributes to the ground state energy of the ionic solid. The important quantity for phonons is how the Coulomb interaction changes when the ion is displaced. The interaction between two ion displacements is a dipole–dipole interaction:

$$\frac{e^2 Z_1 Z_2}{|\vec{R}+\vec{Q}|} = e^2 Z_1 Z_2 \left[\frac{1}{R} - \frac{\vec{Q}\cdot\vec{R}}{R^3} + \frac{1}{2}\boldsymbol{Q}\cdot\boldsymbol{\phi}(\boldsymbol{R})\cdot\boldsymbol{Q} + \cdots\right], \quad (5.59)$$

$$\phi_{\mu\nu}(\vec{R}) = \frac{\delta_{\mu\nu}}{R^3} - \frac{3R_\mu R_\nu}{R^5}. \quad (5.60)$$

The term linear in $\vec{Q}$ can be ignored since its coefficient vanishes in equilibrium. The quadratic term is very important in lattice dynamics. It gives the dipole–dipole interaction between the dipoles of the oscillating ions. Most theories of phonons keep only quadratic terms in the displacement $\vec{Q}$, which is called the *harmonic approximation.*

The PIM also has short-range pair interactions. These are van der Waals interactions if the ions are well separated. If they are near neighbors, and if their electronic charges overlap, there are additional forces due to bonding. The ion displacements $Q_{j\mu}$ are a small fraction of the ion separation. Taylor series can be used to expand the potential in powers of the displacements. Central forces are a function only of the magnitude of the distance between e two ions. Let $\vec{R}_j$ and $\vec{R}_\ell$ be their equilibrium positions, and $\vec{R}_{j\ell}$ be their separation. The force constant is found as follows:

$$V(\vec{R}_{j\ell}+\vec{Q}_{j\ell}) = V(\vec{R}_{j\ell}) + \frac{1}{2}Q_{j\ell,\mu}Q_{j\ell,\nu}K_{\mu\nu}(\vec{R}_{j\ell}) + \ldots, \quad (5.61)$$

$$K_{\mu\nu}(\vec{R}_{j\ell}) = \nabla_\mu\nabla_\nu V(\vec{R}_{j\ell}) = B\delta_{\mu\nu} + \frac{R_\mu R_\nu}{R^2}(A-B), \quad (5.62)$$

$$A = \frac{\partial^2 V}{\partial R^2}, \quad (5.63)$$

$$B = \frac{1}{R}\frac{\partial V}{\partial R}. \quad (5.64)$$

The parameters A and B are usually defined for first and second neighbors. In binary crystals of plus and minus ions, they are usually labeled A_{+-} for first neighbors, and A_{++}, A_{--} for second neighbors. They are found by fitting to the experimental spectra. Usually A_{+-} is much larger in magnitude than B_{+-}, since the potential $V(R)$ is usually near its minimum at the separation of first neighbors, R_1. After the short range parameters, the next important feature of the polarizable ion model is the polarization of the ions. Only dipole polarization is included in the PIM. More elaborate models add additional polarizations. Some of these are discussed below. Each ion j

has an electronic dipole moment $\vec{p}_j$ induced by the local electric field. This moment is given by the product of the ion polarizability $\alpha_{d,j}$ and the local electric field $\vec{\mathcal{E}}_j^{(l)}$. The polarizability is assumed to be a scalar, which is correct for ions with filled electronic shells.

The local electric field at an ion site j is found by adding all of the fields from the other ions. Each ion has two contributions to the dipole moment. One is the electronic moment $\vec{p}_j$ and the other is from the ion displacement $eZ_j\vec{Q}_j$, where Z_j is the ion charge. The total dipole moment at an ion is $\vec{p}_j + eZ_j\vec{Q}_j$. The local electric field is determined by the dipole–dipole interaction from all of these dipoles:

$$\vec{p}_j = \alpha_{dj}\vec{\mathcal{E}}_j^{(l)}, \tag{5.65}$$

$$\vec{\mathcal{E}}_j^{(l)} = \sum_{\ell\neq j}\phi(\vec{R}_{j\ell})\cdot[\vec{p}_\ell + eZ_\ell(\vec{Q}_\ell - \vec{Q}_j)]. \tag{5.66}$$

The dipole–dipole tensor ϕ is given in (5.60). These equations provide a self-consistent method of calculating the electronic moment $\vec{p}_j$. The electronic polarization provides some screening of the ion–ion interaction. These equations are actually a definition of the dielectric function due to the electronic polarization in the dipole approximation.

The forces on the ion are the summation of the pair forces from the first two neighbors plus the dipole–dipole forces from distant ions. These are called the short-range (SR) and long-range (LR) components, respectively. The distances to the first neighbors are denoted $\vec{\delta}$. Then, one has

$$m_j\omega^2 Q_{j\mu} = F_{j\mu}^{LR} + F_{j\mu}^{SR}, \tag{5.67}$$

$$F_{j\mu}^{SR} = \sum_{\vec{\delta}}[B(\vec{\delta})(Q_{j\mu} - Q_{j+\delta,\mu}) + (A(\vec{\delta}) - B(\vec{\delta}))\delta_\mu\hat{\delta}\cdot(\vec{Q}_j - \vec{Q}_{j+\delta})], \tag{5.68}$$

$$F_{j\mu}^{LR} = eZ_j\mathcal{E}_{j\mu}^{(\ell)}. \tag{5.69}$$

These equations are converted to a dynamical matrix using the same technique discussed for (5.35). Both $\vec{p}_j$ and $\vec{Q}_j$ are converted to functions of wave vector $\vec{k}$. In this representation the set of linear equations have a dimensionality of $6L$, where L is the number of ions in the crystal unit cell: $3L$ from the ions and $3L$ from the electronic moments. The PIM is often used to calculate the lattice dynamics of simple ionic solids such as the alkali halides (Cohen *et al.*, 1987). One can vary the short-range force parameters

$A_{\pm,\pm}$ and $B_{\pm,\pm}$ and obtain a reasonable fit of the PIM to the experimental phonon spectra. The major problem with the PIM is that it predicts too large of a difference between the LO and TO phonon frequencies at the Γ point ($\vec{k} = 0$). This difficulty can be eliminated by letting the ion charge Z_j be a variational parameter with noninteger values. Good fits for most alkali halides are obtained for values $Z = 0.7 - 0.8$. These values have caused many workers to conclude that the ions in alkali halides have noninteger charges. That is, there is a significant amount of electronic bonding, which shares the charge between pairs of ions. However, there is much evidence to show that the ions have integer valence. First, the values for the ground state energy, when compared against measurements of cohesive energy, only make sense by assuming integer valence. Also, numerous electronic energy band calculations such as that of Jennison and Kunz (1976) find negligible bonding. Charge density maps from x-ray scattering also show the amount of electronic charge expected from integer valence, as mentioned in Section 5.1.2.

Noninteger values of the effective charge occur in lattice dynamics for a different reason—namely, the polarization of the ions. So far our theory has included only dipole polarization. Other types of polarization need to be taken into account. The polarization is a dynamic contribution which only happens when the ions are displaced from equilibrium. Both the shell model and the deformation-dipole model include these polarization phenomena.

5.5.4 Deformation-Dipole Model

The deformation-dipole (DD) model was invented by Hardy (1959, 1962), and by Hardy and Karo (1963), to account for dynamical polarization. They introduced the deformation-dipole tensor $\gamma_{\mu\nu}$ which characterizes the amount of additional polarization induced on an ion by the displacement of a neighboring ion. The idea is to have a more accurate description of the dynamic dipole moment $\vec{p}_j$ at an ion site $\vec{R}_j$. This dipole moment is due to electronic polarization. The PIM model has this moment given by the ion polarizability times the local electric field in (5.53):

$$\vec{p}_j = \alpha_{d,j}\vec{\mathcal{E}}_j^{(\ell)}.$$

This formula is not accurate for interactions among nearest neighbors. A more accurate description of polarization is needed between ions whose electronic charge distributions overlap.

A nearby atom at site $\vec{R}_\ell$ displaces a distance $\vec{Q}_\ell$. This displacement creates a dynamic dipole moment $\vec{q}_\ell = eZ_\ell\vec{Q}_\ell$. In linear response, there must be a proportionality between $\vec{q}_\ell$ and $\vec{p}_j$.

The local electric field from dipole–dipole interactions is $\vec{\mathcal{E}}_j^{(\ell)} = \phi(\vec{R}_{j\ell})\cdot\vec{q}_\ell$. In the PIM, the proportionality between $\vec{p}_j$ and $\vec{q}_\ell$ is given by using the dipole approximation for the local field. This result is accurate if the ions are well separated in space. When the ions are nearest neighbors, their charge distributions overlap. The dipole model is not adequate to describe the polarization for overlapping charge distributions. So Hardy and Karo defined the deformation-dipole tensor to be the additional polarization which occurs,

$$p_{j,\mu} = [\alpha_{d,j}]_{\mu\lambda}\mathcal{E}_{j\lambda}^{(\ell)} + [\gamma_j(\vec{R}_{j\ell})]_{\mu\nu}eZ_\ell[Q_{\ell\nu} - Q_{j\nu}]. \tag{5.70}$$

The first term is the dipole polarization which is part of the PIM. The second is the additional polarization which occurs when the charge distributions overlap on neighboring ions. It includes higher moments such as quadrupole and octupole, plus quantum effects due to exchange and correlation. The deformation-dipole tensor is dimensionless.

The tensor $\gamma_{\mu\nu}$ usually has appreciable value only for ions which are nearest neighbors. Let $\vec{\delta}$ denote this nearest neighbor distance. The DD tensor is diagonal: It has a longitudinal component γ_ℓ when the vectors $\vec{\delta}$ and $\vec{q}$ are parallel, and a transverse component γ_t when the two vectors are perpendicular. The trace of the tensor is $\gamma_\ell + 2\gamma_t$. A quantity which is useful is the sum of the tensor components over all first neighbors. For cubic crystals with $\mathcal{Z}$ nearest neighbors, define

$$\begin{aligned}\Gamma &= \sum_{\vec{\delta}} \hat{\eta}\cdot\gamma(\vec{\delta})\cdot\hat{\eta}, \\ &= \frac{\mathcal{Z}}{3}[\gamma_\ell + 2\gamma_t],\end{aligned} \tag{5.71}$$

where $\hat{\eta}$ is a unit vector in any direction. The notation $\gamma(\vec{\delta})$ means the DD tensor caused by a displacement of the neighbor at $\vec{\delta}$. The quantity Γ is important in the definition of the effective charge.

It is useful to consider how the deformation-dipole tensor enters into the effective Hamiltonian. In a representation in which the ion displacements $\vec{Q}_j$ and electronic polarization $\vec{p}_j$ are both variables, the DD interaction is provided by a term

$$H_{DD} = -\sum_{j,\vec{\delta}} eZ_{j+\delta}\mathcal{E}_{j\mu}^{(\ell)}\gamma_j(\vec{\delta})_{\mu\nu}[Q_{j,\nu} - Q_{j+\delta,\nu}]. \tag{5.72}$$

The interaction depends upon the difference of the displacements of the two neighbors. If they displace the same amount, in the same direction, then they do not polarize each other. This interaction term makes two contributions to the dynamical matrix. First, it contributes to the electronic moment on the ion at $\vec{R}_j$. The contribution to the induced moment δp_j is found by taking the derivative with respect to the local field at site j

$$\begin{aligned} \delta p_{j\mu} &= -\frac{\delta H_{DD}}{\delta \mathcal{E}^{(\ell)}_{j\mu}} \\ &= \sum_{\vec{\delta}} eZ_{j+\delta}\gamma_j(\vec{\delta})_{\mu\nu}[Q_{j,\nu} - Q_{j+\delta,\nu}]. \end{aligned} \tag{5.73}$$

This term is just the additional polarization discussed above in (5.70).

However, the interaction H_{DD} also contributes a force on the ions, which is denoted F^{DD}. It is found by taking the derivative with respect to the ion displacement. The force on the mth ion is given by

$$\begin{aligned} F^{DD}_{m,\mu} &= -\frac{\delta H_{DD}}{\delta Q_{m\mu}} \\ &= -e\sum_{\vec{\delta}}[Z_m\gamma_{m-\delta}(\vec{\delta})_{\mu\nu}\mathcal{E}^{(\ell)}_{m-\delta,\nu} - Z_{m+\delta}\gamma_m(\vec{\delta})_{\mu\nu}\mathcal{E}^{(\ell)}_{m,\nu}]. \end{aligned} \tag{5.74}$$

The above term is a consequence of Newton's law of action-reaction. If the ion displacements cause polarization on the ions, then polarization causes forces on the ions.

The DD contributes to the dynamical matrix. Equation (5.67) is modified to

$$m_j\omega^2 Q_{j\mu} = F^{LR}_{j\mu} + F^{SR}_{j\mu} + F^{DD}_{j\mu}, \tag{5.75}$$

where the last term is the DD-force given in (5.74). DD also causes an additional term in the expression for the electronic dipole $\vec{p}_j$:

$$p_{j\mu} = \alpha_{dj}\mathcal{E}^{(\ell)}_{j\mu} + e\sum_{\delta}\gamma_{j,\mu\nu}Z_{j+\delta}(Q_{j\nu} - Q_{j+\delta,\nu}). \tag{5.76}$$

The last equation is the same as (5.70). The following expressions are *not* changed: (5.66) for the local field $\mathcal{E}^{(\ell)}_{j\mu}$, (5.68) for the short-range force, and (5.69) for the long-range force. These equations constitute the DD model of lattice dynamics. They are changed to collective coordinates by the Fourier transformation given in (5.54). For a solid with L ions per unit cell, the resulting equations have a dimensionality of $9L$. That can be reduced to $6L$ by eliminating the variable $\vec{p}_j$ by inserting (5.70) into (5.66). The end result is a dynamical matrix of dimension $6L$. This model has been applied to the calculation of lattice dynamics of alkali halides (Hardy and Karo, 1979) and zincblende semiconductors (Kunc *et al.*, 1975). The model provides a good fit to the phonon spectra of binary solids. Mahan (1986) showed that the deformation-dipole parameters (γ_ℓ, γ_t) could be calculated using TDLDA. The numerical method is a simple extension of the method used to calculate the polarizability. Section 3.3 describes how TDLDA can be used to calculate the response of an atom to any external perturbation. An example is provided by the calculation of the DD parameters.

The spherical solid model is used. The first step is to derive an expression for the forces on the atom. There are two forces which are important for the deformation-dipoles. The first is due to the local electric field ($\vec{\mathcal{E}}^{(i)} \equiv \vec{\mathcal{E}}$), which is represented by an interaction potential

$$V_{\mathcal{E}}(\vec{r}) = -e\vec{r}\cdot\vec{\mathcal{E}}. \tag{5.77}$$

This potential energy acts on the electron at position $\vec{r}$ in the atom at $\vec{R}_i$.

The second important force comes from the neighboring ions. Let $V_{ei}(\vec{r}+\vec{R}_{mi})$ be the potential energy between the electron at $\vec{r}$ and the neighboring ion at $\vec{R}_m$. Often this potential is represented by a pseudopotential. When the neighboring ions are in their equilibrium positions, this potential energy contributes to the ground state energy of the crystal. The deformation-dipoles come from a relative displacement $\vec{Q}_{mi} = \vec{Q}_m - \vec{Q}_i$ of the neighbor ion. The displacement causes a change in the neighboring interaction which is found by expanding the potential in a Taylor series,

$$V_{ei}(\vec{r}+\vec{R}_{mi}+\vec{Q}_{mi}) = V_{ei}(\vec{r}+\vec{R}_{mi}) + \vec{Q}_{mi}\cdot\vec{\nabla}V_{ei}(\vec{r}+\vec{R}_{mi}) + \cdots. \tag{5.78}$$

The second perturbation, denoted by V_F, is given by

$$V_F(\vec{r}) = \sum_{m\neq i}\vec{Q}_{mi}\cdot\vec{F}(\vec{r},\vec{R}_{mi}), \tag{5.79}$$

$$\vec{F}(\vec{r},\vec{R}_{mi}) = \vec{\nabla}V_{ei}(\vec{r}+\vec{R}_{mi}). \tag{5.80}$$

The two perturbations are added. Together they change the ground state energy, which is evaluated in the second order of perturbation theory:

$$V' = V_{\mathcal{E}} + V_F, \tag{5.81}$$

$$\delta E_G = -\sum_{\alpha,\beta} \frac{<\alpha|V'|\beta><\beta|V'|\alpha>}{\lambda_\beta - \lambda_\alpha}. \tag{5.82}$$

The summation α is over occupied electron states, while β is over the empty states. Since V' has two terms, the second-order energy has four terms:

1. The term proportional to $V_{\mathcal{E}}^2$ equals $-\frac{1}{2}\alpha_d\mathcal{E}^2$. The second-order expression is (4.14) for the dipole polarizability.

2. The term proportional to V_F^2 is the basis for the I^3 model of Mahan and Mostoller (1986). It is discussed in the next section.

3. The DD interaction comes from the two cross terms between $V_{\mathcal{E}}$ and V_F.

The cross terms contribute to the ground state energy a term which corresponds to H_{DD} of (5.72) in the effective Hamiltonian. After identifying the cross term as H_{DD}, the definition of the deformation-dipole becomes:

$$Z_m\bar{\gamma}_{i,\mu\nu} = \sum_{\alpha,\beta} \frac{<\alpha|r_\mu|\beta><\beta|F_\nu(\vec{r},\vec{R}_{mi})|\alpha> + <\alpha|F_\nu(\vec{r},\vec{R}_{mi})|\beta><\beta|r_\nu|\alpha>}{\lambda_\beta - \lambda_\alpha}. \tag{5.83}$$

Several features of this expression need explanation. First, an overbar is put on γ to indicate that this expression is not exactly the deformation-dipole parameter of Hardy and Karo. Recall that they defined it to be the additional polarization beyond the dipole approximation. The above expression for $\bar{\gamma}$ includes the dipole contribution. It must be subtracted out from $\bar{\gamma}$ to get γ. If the polarizability is isotropic,

$$\gamma_{j,\mu\nu} = \bar{\gamma}_{j,\mu\nu} - \alpha_{d,j}\phi_{\mu\nu}(\vec{r}_{mi}), \tag{5.84}$$

$$\gamma_\ell = \bar{\gamma}_\ell + \frac{2\alpha_d}{R_1^3}, \tag{5.85}$$

$$\gamma_t = \bar{\gamma}_t - \frac{\alpha_d}{R_1^3}, \tag{5.86}$$

where R_1 is the distance to the first neighbor. Separate formulas have been given for the longitudinal and transverse components of the DD tensor.

The second feature of equation (5.83) is the factor of Z_m on the left. Without this factor, the sign of the deformation-dipole depends upon the sign of the charge of the neighboring ion. Adding this dimensionless factor of Z_m makes the deformation-dipoles for all ions have the same sign.

Section 4.1 discussed the evaluation of the linear polarizability of atoms and ions. The numerical method is to calculate the infinitesimal change in the electron density caused by the application of an infinitesimal electric field. This method can be summarized by the following pair of equations:

$$n(\vec{r},\vec{\mathcal{E}}) = n(\vec{r}) + n_{\mu}^{(1,\mathcal{E})}(\vec{r})\mathcal{E}_{\mu} + O(\mathcal{E}^2), \tag{5.87}$$

$$\alpha_{d,\mu\nu} = e^2 \int d^3 r r_{\mu} n_{\nu}^{(1,\mathcal{E})}(\vec{r}). \tag{5.88}$$

The first step is to calculate the first-order change in density $n_{\nu}^{(1,\mathcal{E})}(\vec{r})$ caused by the infinitesimal electric field. The second step is to take the integral of this change in density with the r-dependent part of the perturbation.

The polarizability can also be written, as in (4.14), as a formula derivable from second-order perturbation theory. The evaluation of the second-order expression is equivalent to the alternate method of finding the first-order change in the density.

The equivalence of these two methods is used to evaluate (5.83) for the DD parameters. Rather than summing over the excited states $|\beta >$, we turn the calculation into an evaluation of the first-order change in the electron density. Since there are two perturbations, there are two terms in the first-order change in the particle density:

$$n(\vec{r},\vec{\mathcal{E}},\vec{Q}) = n(\vec{r}) + n_{\mu}^{(1,\mathcal{E})}(\vec{r})\mathcal{E}_{\mu} + n_{\mu}^{(1,Q)}(\vec{r})Q_{\mu} + \ldots, \tag{5.89}$$

$$\bar{\gamma}_{\mu\nu} = \int d^3 r [F_{\mu}(\vec{r}) n_{\nu}^{(1,\mathcal{E})}(\vec{r}) + r_{\mu} n_{\nu}^{(1,Q)}(\vec{r})]. \tag{5.90}$$

The two terms in the integral for the DD parameters are the two terms in the second-order expression (5.83). The change in density caused by the field is correlated with the force F_{μ}. The change in density caused by the force is correlated with r_{μ}.

Chapters 3 and 4 discussed the evaluation of the change in density due to an infinitesimal field. One solves the first-order equation:

$$[H_0 - \lambda_j]\psi_{j,\mu}^{(1,\mathcal{E})}(\vec{r}) = r_{\mu}\psi_j(\vec{r}), \tag{5.91}$$

$$n_{\mu}^{(1,\mathcal{E})}(\vec{r}) = 2\Re\{\sum_j \psi_j^*(\vec{r})\psi_{j,\mu}^{(1,\mathcal{E})}(\vec{r})\}. \tag{5.92}$$

The screening terms are not shown on the right-hand side for simplicity. A similar method is used to obtain the first-order change in density from the force:

$$[H_0 - \lambda_j]\psi_{j,\mu}^{(1,Q)}(\vec{r}) = F_\mu(\vec{r})\psi_j(\vec{r}), \tag{5.93}$$
$$n_\mu^{(1,Q)}(\vec{r}) = 2\Re\{\sum_j \psi_j^*(\vec{r})\psi_{j,\mu}^{(1,Q)}(\vec{r})\}. \tag{5.94}$$

The numerical method of evaluating the first-order response to the force F_μ is identical to the method of evaluating the response to an external electric field. In spherical coordinates the force has the variables (r, θ, ϕ). The angular dependence is given by spherical harmonic functions $Y_{\ell m}(\theta, \phi)$. The force is expanded in a series of these angular functions

$$F_\mu(\vec{r}, \vec{R}) = \sum_{\ell m} f_{\mu,\ell m}(r, R) Y_{\ell m}(\theta, \phi). \tag{5.95}$$

The force depends upon the distance R to the neighboring ion. The change in density from each angular component is evaluated separately. The evaluation of the DD parameters requires the correlation of the force with the vector $\vec{r}$. Since the position vector has angular functions with $\ell = 1$, then only $\ell = 1$ terms are needed in the evaluation of the force. By choosing a coordinate system where the $\hat{z}$ direction is towards the neighbor $\vec{R}$, only $m = 0$ terms are needed in the summation over angular functions. The change in the procedures, compared to the polarizability, is to have the applied field be a function $f(r)$ rather than just r itself. In all cases the applied field is evaluated self-consistently using (3.55). Again this is done by iteration. Table 5.8 shows the results of Mahan (1986) for $\gamma_{\ell,t}$ for some alkali halides. Also shown are some experimental values. The experimental results are obtained by a method suggested by Karo and Hardy (1963), based on fitting the DD parameters to the Szigeti (1950) charge e^*. The values for γ_ℓ are between 0.1 and 0.2, while those for γ_t are between -0.015 and -0.020.

In binary ionic crystals, such as the alkali halides, the difference between the LO and TO phonon frequencies at $\vec{k} = 0$ is caused by long-range Coulomb forces from the oscillating dipoles. In the DD theory, this difference is given by the formula

$$\omega_{LO}^2 - \omega_{TO}^2 = \frac{4\pi e^{*2}}{3\mu V_0 \epsilon}\left(\frac{\epsilon + 2}{\epsilon}\right)^2, \tag{5.96}$$
$$e^*/e = 1 - \Gamma_c - \Gamma_a, \tag{5.97}$$

where μ is the reduced mass of the two ions, ϵ is the electronic part of the dielectric constant at low frequency (sometimes called ϵ_∞), V_0 is the volume

Table 5.8: Deformation-dipole parameters for some alkali halides. Values are dimensionless. First set of values calculated using TDLDA by Mahan (1986). The second set of values are obtained by fitting to experiments, using a procedure due to Karo and Hardy (1963).

	TDLDA		KH	
Salt	γ_ℓ	γ_t	γ_ℓ	γ_t
LiF			0.140	-0.020
LiCl			0.154	-0.019
NaF	0.167	-0.024	0.119	-0.014
NaCl	0.160	-0.018	0.155	-0.018
NaBr	0.160	-0.018	0.172	-0.019
NaI			0.185	-0.020
KF	0.135	-0.017	0.077	-0.008
KCl	0.149	-0.016	0.131	-0.013
KBr	0.151	-0.017	0.155	-0.015
KI			0.173	-0.016
RbF	0.117	-0.014		
RbCl	0.130	-0.014		
RbCl	0.139	-0.016		

of the unit cell in the crystal, and Γ is defined in (5.71). The subscripts on Γ denote the anion and cation. Their deformation-dipole contributions are just added at $\vec{k} = 0$. Every quantity except e^* in (5.96) can be found by experiment. So this equation is used to define an experimental value for the effective charge e^*.

Hardy and Karo assume that the DD is determined by a function:

$$\Lambda(r_0) = m_0 \exp(-r_0/\rho), \tag{5.98}$$

$$\gamma_\ell = -\frac{\partial \Lambda}{\partial r_0} = \Lambda/\rho, \tag{5.99}$$

$$\gamma_t = -\Lambda/r_0, \tag{5.100}$$

$$\Gamma = 2\Lambda(r_0)(1/\rho - 2/r_0). \tag{5.101}$$

The first neighbor distance is $r_0 \equiv R_1$, and ρ is a length parameter which depends upon electron density. In preparing the last two columns of Table 5.8 we used values of e^* from Lowndes and Martin (1969), and values of ρ from Karo and Hardy (1963). Hardy and Karo assumed only the anion was polarizable ($\Gamma_c = 0$), and we followed that assumption in preparing the table of values.

According to the model of Hardy and Karo, the transverse component γ_t must have the opposite sign from γ_ℓ. Since $r_0/\rho \sim 10$, the transverse component is about 8-10 times smaller in magnitude. The theoretical results using TDLDA give similar values. Indeed, the agreement between theory and experiment is very good except for F^-. The TDLDA can be used to calculate the DD parameters for the cations, and they are generally much smaller than the values for anions. The negative ions are much more polarizable. We conclude that the TDLDA method is a successful way to calculate the DD parameters. So far it is the only theoretical method used to calculate these parameters.

The Szigeti charge e^* is an important parameter in lattice dynamics. Often it is interpreted as showing that the ions have a fractional valence. We do not agree with that interpretation. Instead, the ions have an integer valence in ionic solids. However, when an ion is displaced from equilibrium, it polarizes the surrounding ions. This polarization changes the apparent dipole moment caused by the displacement from $eZ_j\vec{Q}_j$ to $e^*Z_j\vec{Q}_j$. The Szigeti charge is a dynamical property. The difference between e^* and e is given by the DD.

5.5.5 Indirect Ionic Interactions

The second-order change in the ground state energy (5.82) has four terms. One is the energy change due to an electric field, and two are related to the deformation-dipole. The fourth term is the second-order energy from V_F. It is the basis for the *indirect ionic interactions* (I^3) of Mahan and Mostoller (1986). Call this term V_{I3} and rewrite it as

$$V_{I3,i} = \sum_{m,j} \bar{M}_{\mu\nu}(\vec{R}_{ij}, \vec{R}_{im}) Q_{\mu,im} Q_{\nu,ij}, \tag{5.102}$$

$$Q_{\mu,ij} = Q_{\mu,i} - Q_{\mu,j}, \tag{5.103}$$

$$\bar{M}_{\mu\nu}(\vec{R}, \vec{R}') = \sum_{\alpha,\beta} \left\{ \frac{< \alpha|F_\mu(\vec{r}, \vec{R})|\beta >< \beta|F_\nu(\vec{r}', \vec{R}')|\alpha >}{\lambda_\beta - \lambda_\alpha} \right.$$

$$+\frac{<\alpha|F_\nu(\vec{r}',\vec{R}')|\beta><\beta|F_\mu(\vec{r},\vec{R})|\alpha>}{\lambda_\beta-\lambda_\alpha}\Bigg\}. \quad (5.104)$$

The central ion is i. There is a potential energy of interaction between the relative displacements of two of its neighbors j and m. The above expression is correct if $j \neq m$, or $\vec{R} \neq \vec{R}'$. There is only one term on the right when $j = m$.

I^3 is an example of a three-body interaction. A displacement of a neighboring ion causes the ion at $\vec{R}_i$ to polarize. This polarization exerts a force on all of the neighboring ions of i. There is even a force on the original ion which caused the polarization. This self-energy adds a contribution to the short-range parameters A_{+-} and B_{+-}. The important phenomenon is that the central ion at $\vec{R}_i$ causes an effective interaction between any pair of its neighbors.

This three-body force is different, and much larger, than the Axilrod-Teller (AT) interaction discussed in Section 4.3.3. In the AT interaction, all three ions mutually polarize. In I^3 only the central ion polarizes. The I^3 interaction vanishes if all the ions are neutral, and are separated at large distance. The AT interaction does not vanish in this limit. The I^3 interaction is not the same as the AT.

Let the ions (j, m) be charged with valence $Z_{j,m}$. For the case that all three ions are well separated, there is an interaction term:

$$V_p(\vec{R}_{im},\vec{R}_{ij}) = -Z_m Z_j e^2 \sum_M \frac{\alpha_{M,i}}{(R_{im}R_{ij})^{M+1}} P_M(\cos\theta). \quad (5.105)$$

This energy term arises from the mutual interaction between m and j through the polarization of the ion i. The electrostatic field from one ion, say m, causes the ion i to polarize; the polarization interacts with the other ion, say j. If all three ions are fixed in space, this three-body interaction contributes to the ground state energy. Let the three ions have infinitesimal displacements. The second-order change in energy gives the tensor $\bar{M}_{\mu\nu}$,

$$\lim_{R_{im},R_{ij}\to\infty} \bar{M}_{\mu\nu} = \nabla_{\mu,j}\nabla_{\nu,m}V_p(\vec{R}_{im},\vec{R}_{ij}). \quad (5.106)$$

At large ionic separations, the I^3 interaction is just the classical interaction V_p caused by the electrostatic multipoles induced on ion i. In solids, the neighboring ions are in contact, and this classical expression is not accurate. It is shown here to illustrate the nature of the I^3 interaction.

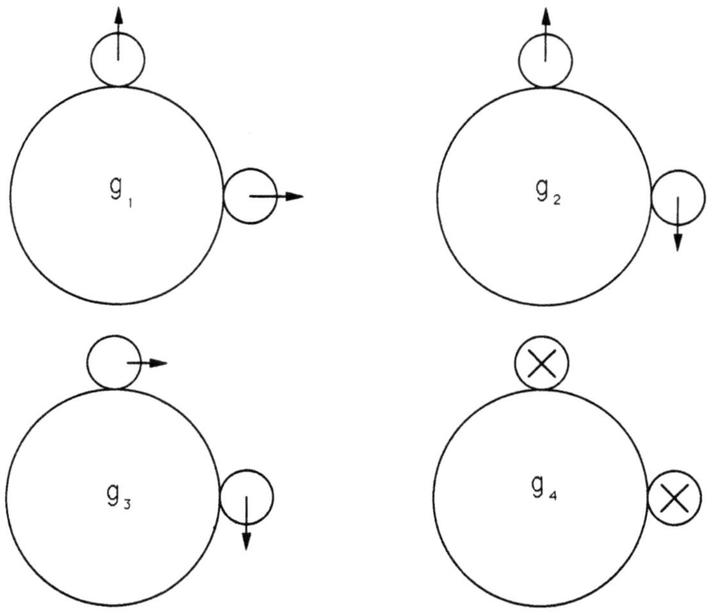

Figure 5.7: The four coupling constants $G_n(\cos\theta)$.

The tensor $\bar{M}_{\mu\nu}$ has nine components. When all three ions are spherically symmetric, symmetry can be used to reduce the number of nonzero components. It is useful to represent the relative displacements of the ions by unit vectors $(\hat{r}, \hat{t}, \hat{n})$ which are in the direction of radial motion $(\hat{r})$, tangential motion $(\hat{t})$, and motion normal to the plane $(\hat{n})$. Nine possible parameters $\bar{G}_n$ can be defined by taking all combinations of three unit vectors for the two pairs of displacements. The overbar on G is explained below. If $\hat{\xi}$ denotes one of the three unit vectors, then define $\bar{G}_n = \hat{\xi} \cdot \mathbf{M} \cdot \hat{\xi}'$. The unprimed vector $\hat{\xi}$ denotes the relative displacement of one ion, say m, while the primed vector $\hat{\xi}'$ denotes the displacement of the other neighbor, say j. If both neighbors are at the same distance R, only four of the G_n are nonzero by symmetry. These are illustrated in Figure 5.7 The only two parameters are the neighbor distance R and the bond angle $\beta \equiv \cos\theta$:

$$\bar{G}_1(R,\beta) = \hat{r} \cdot \mathbf{M} \cdot \hat{r}', \tag{5.107}$$

$$\bar{G}_2(R,\beta) = \hat{r} \cdot \mathbf{M} \cdot \hat{t}', \tag{5.108}$$

$$\bar{G}_3(R,\beta) = \hat{t} \cdot \mathbf{M} \cdot \hat{t}', \tag{5.109}$$

$$\bar{G}_4(R,\beta) = \hat{n} \cdot \mathbf{M} \cdot \hat{n}'. \tag{5.110}$$

The nine components of the tensor $\hat{M}_{\mu\nu}$ have been reduced to four nonzero constants. The others vanish by symmetry. Values of $\bar{G}_n$ with only one factor of $\hat{n}$, such as $\hat{n} \cdot \mathbf{M} \cdot \hat{r}'$, vanish by reflection symmetry through the plane of the three atoms.

The coupling constants $\bar{G}_n$ depend upon the bond angle $\beta = \cos\theta$. They can be expressed as a series over angular momentum ℓ. Each term in the series has an angular function $\mathcal{A}_\ell$ times a function $B_\ell(R)$. The easiest method of finding the angular functions is to evaluate $(\hat{\xi} \cdot \vec{\nabla})(\hat{\xi}' \cdot \vec{\nabla}')V_p$ in (5.105). That gives the correct result at large values of R. The angular functions $\mathcal{A}_\ell$ do not depend upon R, and are correct at all values of distance. At large values of R the coupling constants are:

$$\bar{G}_1 = -(Ze)^2 \sum_\ell (\ell+1)^2 \frac{\alpha_\ell}{R^{2\ell+4}} P_\ell(\beta), \tag{5.111}$$

$$\bar{G}_2 = (Ze)^2 \sum_\ell (\ell+1) \frac{\alpha_\ell}{R^{2\ell+4}} P_\ell^1(\beta), \tag{5.112}$$

$$\bar{G}_3 = (Ze)^2 \sum_\ell \frac{\alpha_\ell}{R^{2\ell+4}} \left[P_\ell^2 - \beta \frac{\partial P_\ell}{\partial \beta} \right], \tag{5.113}$$

$$\bar{G}_4 = -(Ze) \sum_\ell \frac{\alpha_\ell}{R^{2\ell+4}} \frac{\partial P_\ell}{\partial \beta}. \tag{5.114}$$

The superscripts on the Legendre polynomials refer to the associated Legendre polynomials.

The coupling constants $\bar{G}_n$ are evaluated using TDLDA. The previous section discussed the method of evaluating the change in density $n_\mu^{(1,Q)}(\vec{r}, \vec{R})$ caused by an infinitesimal displacement Q_μ of a neighbor at $\vec{R}$. The I^3 force tensor is found using

$$M_{\mu\nu}(\vec{R}, \vec{R}') = \int d^3r [F_\mu(\vec{r}, \vec{R}) n_\nu^{(1,Q)}(\vec{r}, \vec{R}') + F_\nu(\vec{r}, \vec{R}') n_\mu^{(1,Q)}(\vec{r}, \vec{R})]. \tag{5.115}$$

The change in density from one ion displacement is correlated with the force from the other ion displacement. These forces depend upon the bond angle $\beta = \cos(\hat{R} \cdot \hat{R}')$ between $\vec{R}$ and $\vec{R}'$. The angular dependence is expressed by expanding the function in a series over the angular momentum. One way to proceed is to start with the electron-ion interaction, and expand it:

$$V_{ei}(\vec{r} + \vec{R}) = \sum_\ell V_\ell(r, R) P_\ell(\cos\theta), \tag{5.116}$$

$$V_\ell(r, R) = \frac{2\ell+1}{2} \int_0^\pi d\theta \sin\theta P_\ell(\cos\theta) V_{ei}(\vec{r} + \vec{R}), \tag{5.117}$$

where θ is the angle between the vectors $\vec{r}$ and $\vec{R}$. The force is the gradient of this interaction. In order to find $\bar{G}_n$ we need the projection of this force on our three unit vectors,

$$\hat{\xi}\cdot\mathbf{F} = \hat{\xi}\cdot\vec{\nabla}V_{ei}. \tag{5.118}$$

If $\hat{\xi} = \hat{r}$, the derivative is taken with respect to R. For $\hat{\xi} = (\hat{t},\hat{n})$ the derivative is taken with respect to the angular functions. For each value of angular momentum, there are two functions which are important:

$$f_\ell(r,R) = \frac{\partial}{\partial R}V_\ell(r,R), \tag{5.119}$$

$$v_\ell(r,R) = \frac{1}{R}V_\ell(r,R). \tag{5.120}$$

These two radial functions are used in the modified Sternheimer method. The notation $< v|f >_\ell$ is the sum of two terms: one is to calculate the change in density from the perturbation f_ℓ and then correlate the result with the perturbation v_ℓ; the second term is to calculate the change in density from v_ℓ and then to correlate the result with f_ℓ. In terms of this notation, the final expression for the coupling coefficients are:

$$\bar{G}_1 = \sum_\ell P_\ell(\beta) < f|f >_\ell, \tag{5.121}$$

$$\bar{G}_2 = \sum_\ell P_\ell^1(\beta) < f|v >_\ell, \tag{5.122}$$

$$\bar{G}_3 = -\sum_\ell \left[P_\ell^2 - \beta\frac{\partial P_\ell}{\partial\beta}\right] < v|v >_\ell, \tag{5.123}$$

$$\bar{G}_4 = \sum_\ell \frac{\partial P_\ell}{\partial\beta} < v|v >_\ell . \tag{5.124}$$

TDLDA is used to calculate the coefficients $< f|f >_\ell$, etc. Table 5.9 shows values of $< f|f >_\ell$ for K^+ and Br^- in KBr, and Ag^+ in AgBr, for various values of ℓ. The natural units for force constants in lattice dynamics are e^2/V_0, where V_0 is the volume per unit cell. The numerical results fall off slowly with increasing value of angular momentum ℓ. The values for K^+ are somewhat larger than those for Br^-, while those for Ag^+ are the largest. They do not correlate with the polarizabilities of the ions. The polarizability of bromine is the largest, followed by silver and potassium. The numerical values are large, particularly for silver. The I^3 interaction makes an important contribution to the dynamical matrix.

Table 5.9: The correlation integral $- < f|f >_\ell$ in units of e^2/V_0 as a function of angular momentum ℓ. The second and third columns give values for K^+ and Br^- in KBr. The last column is for Ag^+ in AgBr.

ℓ	K^+	Br^-	Ag^+
0	0.11	0.18	0.25
1	0.89	0.42	2.75
2	0.57	0.23	2.12
3	0.42	0.23	0.85
4	0.21	0.23	0.35
5	0.13	0.17	0.30
6	0.09	0.12	0.25
7	0.07	0.06	0.17
8	0.05	0.03	0.12

The I^3 interaction is included in the dynamical matrix using the interaction (5.102). The summation over (j, m) is restricted to the first neighbors of the central ion i. The I^3 tensor $\bar{M}_{\mu\nu}$ is written with a bar over the M. This notation follows that of the DD tensor $\bar{\gamma}_{\mu\nu}$. The overbar denotes the fundamental interaction. For the DD tensor, we subtracted out the dipolar contribution. The remaining DD tensor contained the pair polarization beyond the dipole approximation. The same renormalization must be done for the I^3 tensor $\bar{M}$. It contains the dipole–dipole interaction, which must be subtracted in order to avoid counting this interaction twice. Similarly, the tensor $\bar{M}$ contains some contributions from DD, which must also be subtracted to avoid double counting. Thus we define a renormalized interaction as

$$\begin{aligned} M_{\mu\nu} &= \bar{M}_{\mu\nu} + (Ze)^2[\phi_{\mu\lambda}(\vec{R})\alpha_{d,\lambda\iota}\phi_{\iota\nu}(\vec{R}') \\ &\quad + \phi_{\mu\lambda}(\vec{R})\gamma_{\lambda\nu}(\vec{R}') + \gamma_{\mu\lambda}(\vec{R})\phi_{\lambda\nu}(\vec{R}')], \qquad (5.125) \\ G_1 &= \bar{G}_1 + 4\cos(\theta)\frac{(Ze)^2}{R^3}\left[\frac{\alpha_d}{R^3} - \gamma_\ell\right], \qquad (5.126) \\ G_2 &= \bar{G}_2 - \sin(\theta)\frac{(Ze)^2}{R^3}\left[\frac{2\alpha_d}{R^3} + 2\gamma_t - \gamma_\ell\right], \qquad (5.127) \\ G_3 &= \bar{G}_3 - \cos(\theta)\frac{(Ze)^2}{R^3}\left[\frac{\alpha_d}{R^3} + 2\gamma_t\right], \qquad (5.128) \end{aligned}$$

$$G_4 = \bar{G}_4 + \frac{(Ze)^2}{R^3}\left[\frac{\alpha_d}{R^3} + 2\gamma_t\right]. \tag{5.129}$$

The dimension of G_n is that of a force, and is usually expressed in terms of e^2/V_0. For fcc lattices, $V_0 = 2R_1^3$ if R_1 is the distance to the first neighbor.

For alkali halides, each ion has six first neighbors. The neighbors have bond angles of either $\theta = \pi/2$ or π. The coupling constants with bond angles of $\pi/2$ are denoted $g_n = G_n(\theta = \pi/2)$, and those with bond angles of π are denoted h_n. There are only two nonzero constants at π:

$$h_1 = G_1(\theta = \pi), \tag{5.130}$$

$$h_2 = G_4(\theta = \pi) = G_3(\theta = \pi). \tag{5.131}$$

Mahan and Mostoller (1986, 1990) used TDLDA to calculate the six coupling constants $(g_1, g_2, g_3, g_4, h_1, h_2)$ for each ion in a number of alkali, copper, and silver halides. Some of their results are shown in Table 5.10. Several trends are apparent from this table. First, the values for g_1 and h_1 are usually the largest. These impart radial forces, while the others impart tangential forces. The latter are obviously weaker. The values for g_4 and h_2 can be neglected except perhaps for silver. The important parameters are g_1 and h_1. They should be included in the dynamical matrix for the phonon frequencies of the alkali halides.

Secondly, the values for Ag^+ are much larger than for the other ions. The values for the halides are larger than for the alkalis. However, the values for silver are by far the largest. Similar large values are found for the copper ion in the copper halides with the zincblende structure. The phonon dispersion relations in the copper and silver halides are very different than in the alkali halides (Bilz and Kress, 1979). This difference is due to the large I^3 interactions in the silver and copper salts.

Bilz and collaborators (Fischer *et al.*, 1972) introduced the concept of *quadrupolar deformability* into the lattice dynamics of the copper and silver halides. This concept is similar to I^3. Both depend upon an indirect interaction of two ions by the polarization of a common neighbor. They speculated that silver and copper would have a large value of quadrupolar deformability (QD) because of their filled d-shells. Our calculations have confirmed this hypothesis, since the I^3 parameters are large because of the d-shells. Although QD and I^3 are qualitatively similar, they differ in important details. The I^3 interaction is not quadrupolar — it depends upon a large range of angular momentum values, as shown in Table 5.9. Also , their classification of coupling constants differs from ours, since they omit g_2. Also, we do not

Table 5.10: I^3 coupling constants calculated using TDLDA in units of e^2/V_0.

Ion	Salt	g_1	g_2	g_3	g_4	h_1	h_2
F^-	NaF	-0.14	0.12	0.02	-0.02	0.53	-0.06
	KF	-0.12	0.09	0.01	-0.01	0.45	0.00
	RbF	-0.11	0.08	0.01	-0.01	0.37	0.00
Cl^-	NaCl	-0.11	0.09	0.07	0.00	0.35	0.05
	KCl	-0.12	0.09	0.04	-0.01	0.39	0.03
	RbCl	-0.11	0.08	0.04	-0.01	0.33	0.03
Br^-	NaBr	-0.09	0.08	0.09	0.00	0.30	0.06
	KBr	-0.11	0.09	0.06	-0.01	0.36	0.04
	RbBr	-0.10	0.08	0.05	-0.01	0.33	0.03
Na^+	NaF	0.01	0.00	0.01	0.00	-0.02	0.01
	NaCl	0.03	-0.01	0.01	0.00	-0.03	0.01
	NaBr	0.03	0.00	0.01	0.00	-0.02	0.01
K^+	KF	0.06	-0.02	0.03	0.00	-0.09	0.02
	KCl	0.12	-0.02	0.02	0.00	-0.08	0.02
	KBr	0.14	-0.02	0.02	0.00	-0.05	0.02
Rb^+	RbF	0.10	-0.04	0.05	0.01	-0.13	0.04
	RbCl	0.17	-0.03	0.04	0.01	-0.10	0.02
	RbBr	0.22	-0.04	0.04	0.01	-0.04	0.03
Ag^+	AgF	0.09	0.00	0.15	0.01	-0.10	0.10
	AgCl	0.49	-0.05	0.18	0.02	-0.51	0.10
	AgBr	0.73	-0.03	0.20	0.02	-0.55	0.10

like the word 'deformable,' since it conveys the wrong physical picture. Ions which are deformable do not transfer the force to the other neighbor — it is the rigid ions which do it. Nevertheless, there are numerous qualitative similarities between QD and I^3. For the present discussion, the important point is that these parameters can be calculated using TDLDA. One can develop a model of lattice dynamics in insulators where all of the parameters have a physical interpretation, and can be computed using these simple methods.

Although the I^3 model is physically attractive, its usefulness depends upon whether it describes well the force constant corresponding to three-body forces in ionic solids. This is determined by fitting the model to the phonon spectra of insulators. Mahan and Mostoller (1990) fitted the D^2I^3 model to the phonon spectra of four crystals: KCl, KBr, RbCl, and AgBr. The agreement between the fitted and calculated parameters was qualitative. The parameters which TDLDA predicted to be small were found to be small. The few constants which TDLDA predicted to be large were found to be the most important ones. The parameters were much larger for the Ag^+ ion than for the other ions. The detailed agreement was mixed: some parameters agreed well while others did not agree. More work needs to be done in fitting the model to the phonon spectra of other ionic solids.

5.6 Hyperpolarizability of Alkali Halides

Nonlinear optical susceptibilities enter in the description of a variety of phenomena in solids, such as four-wave mixing, third harmonic generation, coherent- and hyper-Raman spectroscopy, etc. Since the initial experiments following the development of the laser, experimental techniques have steadily improved, and in some cases even the electronic and lattice contributions to nonlinear indices have been determined (see Hellwarth, 1977). However, theoretical calculations of nonlinear susceptibilities have lagged behind. Most of the early theoretical work on nonlinear optical susceptibilities of solids dealt with the simplifications provided by crystal symmetry; e.g., the vanishing of certain components of the susceptibility tensors, and various ratios among the nonzero components. Only in recent years has it been possible to compute absolute magnitudes of nonlinear susceptibilities with any reliability.

The analysis of nonlinear optical processes in dense systems is complicated by competing effects which are not easy to separate. For instance, in polar solids the Raman scattering cross section has the so-called elasto-optic and electro-optic contributions, and it is only in a few cases where an appropriate resonance exists that the relative strength of the two has been assessed. First-principles calculations are needed to test the validity of phenomenological theories, especially to evaluate the magnitude of various contributions. For example, a recent first-principles calculation leads to the conclusion that the dipole-quadrupole susceptibility is more important than the dipole hyperpolarizability in the electro-optic contribution to Raman scattering in the alkali halides (Mahan and Subbaswamy, 1986). Raman scattering is treated in the next section.

As seen in Section 4.4, LDA provides a simple computational scheme for the calculation of nonlinear susceptibilities with the same degree of accuracy as Hartree-Fock based methods. These calculations yield results at the level of semi-quantitative to quantitative agreement with experiments for rare gases. For crystalline solids, however, the calculations have been limited to simplified bond charge models (see, e.g., Levine, 1973), with a notable exception (Baroni and Resta, 1986). Recently there has been a cluster type calculation of the hyperpolarizability of LiF and LiCl within a CHF-MP perturbation scheme (Fowler and Madden, 1985; 1986). A TDLDA calculation of the static electronic hyperpolarizabilities of alkali halide crystals was carried out by Johnson *et al.* (1987).

The justification for the use of LDA for negative ions in crystals has already been discussed in this chapter. In ionic crystals the Madelung potential confines the extra electron to the vicinity of the anion nucleus, giving rise to binding even without self-interaction correction (SIC).

The nonlinear polarizabilities γ_{ijkl} and B_{ijkl} have already been introduced earlier. In a cubic environment there are only two independent components of each of these tensors: $\gamma_{zzzz}, \gamma_{xxzz}, B_{zzzz}$, and B_{xxzz}.

The LDA calculation of the ground state in the spherical solid approximation has been discussed in Section 5.1. The neighbor's interactions are represented by pseudopotentials. In the calculation of the nonlinear polarizability in the crystal there are two complications: the exact zero-field density $n_i^{(0)}$ is of course not spherically symmetric; and additionally, an electric field perturbs not just the central ion but neighboring ions as well. If we treat the neighbors as spherical pseudopotentials, the latter complication does not arise. For a cubic crystal, the lowest nonvanishing distortion of the spherically symmetric charge distribution corresponds to $\ell = 4$ (see (5.16)). For the linear polarizability, the cubic distortion makes no contribution. If an electric field is switched on, the first-order induced density $n_i^{(1)}$ will have an

$\ell = 1$ component coming from adding the $\ell = 1$ perturbation to the $\ell = 0$ piece of $n_i^{(0)}$, plus ℓ=3,5, etc., components from the $\ell = 4$ and higher pieces of $n_i^{(0)}$. Hence the $\ell = 1$ component of $n_i^{(1)}$ comes only from the spherically symmetric (ℓ=0) piece of $n_i^{(0)}$. Since the linear polarizability α depends only on this ℓ=1 piece of $n_i^{(1)}$, the cubic fields do not affect α.

This argument cannot be extended to the nonlinear polarizabilities. For example, the dipole hyperpolarizability γ depends on the ℓ=1 component of $n_i^{(3)}$. Following a sequence as above, this term comes from the ℓ=0 and ℓ=2 components of $n_i^{(2)}$, which in turn arise from the ℓ=1 and ℓ=3 pieces of $n_i^{(1)}$, which depend on the ℓ=0 and ℓ=4 components of $n_i^{(0)}$. Nevertheless, we expect the dominant contribution to the hyperpolarizability to come from distortions of the central ion's ℓ=0 zero-field density, and so we spherically average the crystal potential in calculating the diagonal component γ_{zzzz}.

5.6.1 Spherical Solid Model

Johnson *et al.* (1987) have calculated the dipole and quadrupole hyperpolarizabilities $\gamma^{\pm}$ and $B^{\pm}$ for all of the alkali halides with the rock salt structure. They used the spherical solid model in the pseudopotential approach described in Section 5.1, and the Ceperley-Alder correlation energy. To demonstrate the consequence of electron-electron overlap in the crystal they calculated the ionic hyperpolarizabilities in KCl in several situations, and these results are shown in Table 5.11. In the first line of Table 5.11 the polarizabilities of an isolated K^+ cation are included for comparison with the in-crystal results. (The free anion is not shown because in the absence of self-interaction corrections, which was neglected, the isolated anion does not bind.) Line 2 shows the polarizabilities of the (decoupled) K^+ and Cl^- ions placed in a crystal of fixed point charges; i.e., in this calculation the crystal potential is simply the Madelung potential. Rows 3 and 4 are the polarizabilities calculated using the pseudopotential approach, including first-neighbor and first- and second-neighbor shells, respectively.

The first point to note from Table 5.11 is that the polarizabilities — linear and nonlinear — of the cation are essentially unchanged when it is placed in the crystal, while the polarizabilities of the anion depend strongly on its local environment, as already pointed out in Section 5.2. The cation is quite tightly bound, and surrounding one by a shell of point charges changes its electronic density only slightly. For example, surrounding a free K^+ by a lattice of point charges (the Madelung potential) increases γ^+ and B^+ slightly

Table 5.11: Effect of environment on ionic polarizabilities for KCl. "Madelung only" refers to the neglect of all electron overlap effects. The next two rows show the effect of including overlap from the first shell and from the first two shells of neighbors, respectively, in a pseudopotential representation. From Johnson *et al.* (1987).

	α (10^{-24} esu)		$\gamma/6$ (10^{-39} esu)		$-B$ (10^{-40} esu)	
	(+)	(−)	(+)	(−)	(+)	(−)
Free ion	0.850		11.3		14.4	
Madelung only	0.850	5.01	11.8	1810	14.5	881
1 shell	0.838	3.55	10.6	564	13.7	339
2 shells	0.838	3.50	10.4	487	13.6	317

(lines 1 and 2 in Table 5.11), reflecting the expected small expansion of the cation's electron cloud. When electrons are included on the nearest-neighbor shell or two of ions (rather than treating them as point charges), electron-electron overlap repulsion compresses the central ion's electron cloud, leading to a decrease in its polarizabilities. This can be seen in the change (from line 2 to line 3 of Table 5.11) that results when overlap with only the nearest-neighbor shell is included. The cation's polarizabilities are only slightly diminished, but the effect on the anion is quite dramatic. In KCl, the anion's linear polarizability α^- decreases by 32%, and the hyperpolarizability γ^- drops by a factor of 4. These large decreases arise because the electrons are very loosely bound on the negatively-charged anions, and the addition of overlap repulsion substantially increases their binding. From Table 5.11 we see that including next-nearest neighbors changes α^- only slightly (1%), but γ^- changes much more (14%). Thus, the hyperpolarizability is much more sensitive than is the linear polarizability to the tail of the electronic cloud, which is to say the region of overlap.

Table 5.12 shows the hyperpolarizabilities γ and B calculated using four nearest-neighbor pseudopotential shells (with point charges for more distant ions) in the spherical solid model. The trends evident in the linear polarizabilities can also be seen in the pseudopotential calculation of hyperpolarizabilities. First, a cation's hyperpolarizabilities are largely independent of the crystal environment and hence do not change as a cation is paired with different anions. Second, an anion's hyperpolarizabilities γ^- and B^-

Table 5.12: TDLDA hyperpolarizabilities of alkali halides in the spherical solid model. From Johnson *et al.* (1987).

	$\gamma/6$ (10^{-39} esu)		B (10^{-40} esu)	
	(+)	(−)	(+)	(−)
LiF	0.029	40.6	0.073	29.5
LiCl	0.029	201	0.074	167
LiBr	0.029	3.39	0.074	282
LiI	0.029	6.30	0.074	528
NaF	0.602	86.3	1.01	57.9
NaCl	0.593	339	1.00	252
NaBr	0.593	538	1.00	405
NaI	0.593	965	1.00	731
KF	10.4	1.20	13.6	80.1
KCl	10.4	459	13.6	312
KBr	10.3	698	13.5	483
KI	10.2	1210	13.5	850
RbF	29.0	169	34.0	97.9
RbCl	29.0	574	33.9	364
RbBr	28.7	870	33.8	562
RbI	28.4	1470	33.7	966

increase monotonically with the lattice constant, for the same reason given above for the linear case. In fact, the increase is much more marked for hyperpolarizabilities: γ^- increases by a factor of 3 or 4 when Li^+ is replaced by Rb^+, while α^- increases by only 20-60%. Additionally, even more than in the case of the linear polarizability, the total dipole hyperpolarizability $\gamma^+ + \gamma^-$ per ion pair is dominated by the anion's contribution.

Before comparing the calculated values with experiments, we discuss the calculation of the cubic corrections in perturbation theory.

5.6.2 Cubic Anisotropy

As mentioned earlier, for a spherically symmetric system, there is only one independent component of γ_{ijkl}, namely, $\gamma_{1111} \equiv \gamma_1$. In a cubic environment, there is another independent component, $\gamma_{1122} \equiv \gamma_2$. To calculate the latter, one must go beyond the spherical averaging, and include the cubic potential as a correction. Modifications to TDLDA theory described earlier needed to find the cubic correction terms are described below. Note that the methods are similar to those used in the calculation of the deformation-dipole coefficients in Section 5.4. A significant difference is that here we are solving for a fourth-rank tensor quantity. We follow closely the work of Johnson and Subbaswamy (1989).

First, one needs an electric field that has components along two different crystal axes. Assume that the ion's in-crystal ground state electronic density $n(\vec{r})$ has been found in LDA in the absence of an electric field. When a uniform static electric field

$$\vec{\mathcal{E}} = f_x\hat{x} + f_z\hat{z} \tag{5.132}$$

is applied, the electrons feel a (bare) perturbing potential

$$V_F(\vec{r}) = \vec{\mathcal{E}} \cdot \vec{r}, \tag{5.133}$$

where the field $\mathcal{E}$ is in units of $e/2a_0^2$ and energies are in Rydbergs and lengths in Bohr radii (a_0) as before. All induced quantities (e.g., the density or energy) can be expanded in separate powers of the field components f_x and f_z; for instance, the energy is

$$E = E^{(00)} + f_xE^{(10)} + f_zE^{(01)} + f_x^2E^{(20)} + f_z^2E^{(02)} + f_xf_zE^{(11)} + \cdots. \tag{5.134}$$

The hyperpolarizabilities can be written directly as integrals over induced densities or, more conveniently, they can be related to the perturbed energies; for example, the energy fourth order in the field is

$$E = -\frac{1}{24}\gamma_{ijkl}f_if_jf_kf_l; \tag{5.135}$$

so, with a field of the form (5.132), one has

$$\gamma_{1111} = -24E^{(40)} \tag{5.136}$$

$$\gamma_{1122} = -4E^{(22)} \tag{5.137}$$

with γ in units of a_0^7/e^2. Here a superscript (ij) means a term in the energy expansion (5.134) of order $f_x^if_z^j$.

Crystal Potential

In a cubic lattice, the ground state crystal potential can always be written as a sum over kubic harmonics:

$$V_c(\hat{\mathbf{r}}) = \sum_{l=0,4,6,\ldots} v_{cl}(r)\alpha_l(\hat{\mathbf{r}}) \tag{5.138}$$

where the order-l kubic harmonic α_l is a particular sum over spherical harmonics Y_l^m which is invariant under all operations of the octahedral group O_h (von der Lage and Bethe, 1947) and is normalized to unity. For example, the $l = 4$ kubic harmonic is

$$\alpha_4(\hat{\mathbf{r}}) = \sqrt{\frac{7}{12}} Y_4^0(\hat{\mathbf{r}}) + \sqrt{\frac{5}{24}} \left(Y_4^4(\hat{\mathbf{r}}) + Y_4^{-4}(\hat{\mathbf{r}}) \right) \tag{5.139}$$

$$= \frac{5}{8}\sqrt{\frac{21}{\pi}} \left(x^4 + y^4 + z^4 - \frac{3}{5} r^4 \right) \frac{1}{r^4}. \tag{5.140}$$

We will need the following relationship between $\alpha_l(\hat{\mathbf{r}})$ and a sum over Legendre polynomials P_l referred to different axes. Let $\{\hat{\mathbf{R}}\}$ be any set of N_R unit vectors $\hat{\mathbf{R}}$ which transform among themselves under the operations of the octahedral group O_h [e.g., $\{\hat{\mathbf{R}}\}$ could be the $N_R = 6$ vectors of the form $(\pm 1, 0, 0)$]; then

$$\sum_{\hat{\mathbf{R}}} P_l(\hat{\mathbf{r}} \cdot \hat{\mathbf{R}}) = \frac{4\pi}{2l+1} N_R \alpha_l(\hat{\mathbf{r}}) \alpha_l(\hat{\mathbf{R}}), \tag{5.141}$$

where on the right-hand side of (5.141) $\hat{\mathbf{R}}$ is any member of the set $\{\hat{\mathbf{R}}\}$. Obviously the sum in (5.141) is invariant under O_h; hence it can be expanded as a sum over $\alpha_{l'}(\hat{\mathbf{r}})$. Writing each $\alpha_{l'}$ in terms of $Y_{l'}^m$'s, (5.141) follows.

Suppose that one particular ion (either a cation or an anion in the alkali halides) is placed at the center of coordinates. We represent the crystal potential V_c by a sum over pseudopotentials:

$$V_c(\mathbf{r}) = \sum_{\vec{R} \neq 0} V_R(|\vec{r} - \vec{R}|), \tag{5.142}$$

where $\{\vec{R}\}$ is the set of equilibrium ion positions. The ions in a shell at a radius R are all of the same type. Moreover, their position vectors $\vec{R}$ form one or more sets which transform among themselves under O_h. Hence, using (5.138) and (5.141),

$$v_{cl}(r) = \sum_R{}' \lambda_{Rl} v_{Rl}(r), \tag{5.143}$$

where $v_{Rl}(r)$ comes from the expansion of $V_R(|\vec{r}-\vec{R}|)$ over Legendre polynomials oriented along $\vec{R}$:

$$v_{Rl}(r) = \frac{2l+1}{2}\int_{-1}^{1} d(\hat{\mathbf{r}}\cdot\hat{\mathbf{R}})V_R(|\vec{r}-\vec{R}|)P_l(\hat{\mathbf{r}}\cdot\hat{\mathbf{R}}). \tag{5.144}$$

In (5.143)

$$\lambda_{Rl} = \frac{4\pi}{2l+1}N_R\alpha_l(\hat{\mathbf{R}}), \tag{5.145}$$

where N_R is the number of sites in shell R which transform among themselves under O_h and $\alpha_l(\hat{\mathbf{R}})$ is evaluated at any of these sites. The prime in (5.143) is a reminder that ions at a given radius can separate into more than one contribution to the sum.

Perturbation Expansion in the Cubic Potential and the Electric Field

The crystal potential (5.138) has a spherical ($l = 0$) term plus a sum over terms with $l \neq 0$ possessing true cubic symmetry. The former piece can be handled exactly in the LDA calculation of the zero-field ground state for closed-shell ions. Here we incorporate the $l \neq 0$ pieces of V_c to first order in perturbation theory. To this order each value of l contributes additively and can be treated separately (in fact, as will be shown below, only $l = 4$ contributes to γ). Then, in the presence of an electric field (5.132), the unperturbed spherical ground state feels a bare triple perturbation

$$\delta V = \mathcal{E}\cdot\vec{r} + g v_{cl}(r)\alpha_l(\hat{\mathbf{r}}), \tag{5.146}$$

where g is a perturbation parameter which later will be set to unity. The Kohn-Sham equations are formally expanded in f_x, f_z, and g; this is very similar to the expansion in f_x and f_z mentioned above. If in the perturbation expansion of any quantity the term of order $f_x^i f_z^j g^k$ is labeled by a superscript (ijk), then, to first order in g, the hyperpolarizability arises from the energy terms $E^{(041)}$ and $E^{(221)}$ (cf. (5.136), (5.137)).

From (5.141), the cubic piece in (5.146) can be expanded as a sum over P_l's with any arbitrary choice of cubic vectors $\hat{\mathbf{R}}$. This freedom in the choice of cubic vectors simplifies the problem. Since terms first order in g are additive, one can compute the result of each P_l's perturbation separately and combine them at the end. That is,

$$E^{(\alpha\beta 1)} = \frac{1}{\lambda_{Rl}}\sum_{\hat{\mathbf{R}}} U_{\hat{\mathbf{R}}}^{(\alpha\beta 1)}, \tag{5.147}$$

where $U_{\hat{\mathbf{R}}}^{(\alpha\beta 1)}$ is the term of order $f_x^\alpha f_z^\beta g$ arising from a (bare) perturbation

$$\delta V_{\hat{\mathbf{R}}} = \mathbf{f}\cdot\mathbf{r} + g v_{cl}(r) P_l(\hat{\mathbf{r}}\cdot\hat{\mathbf{R}}). \tag{5.148}$$

In (5.148) we can rotate coordinates $\mathbf{r} \to \mathbf{r}'$ so that $\hat{\mathbf{R}}$ is along $\hat{\mathbf{z}}'$ and the electric field again lies in the $x'z'$ plane; in the new coordinates,

$$\delta V_{\hat{\mathbf{R}}} \to \delta V' = \mathbf{f}'\cdot\mathbf{r}' + g V_{cl}(r') P_l(\hat{\mathbf{r}}'\cdot\hat{\mathbf{z}}') \tag{5.149}$$

where

$$\begin{aligned} f_x' &= \sqrt{f_x^2(1-R_x^2) + f_z^2(1-R_z^2) - 2f_x f_z R_x R_z} \\ f_z' &= f_x R_x + f_z R_z. \end{aligned} \tag{5.150}$$

Thus we need to solve (once) the auxiliary problem of the perturbation given by $\delta V'$ in (5.149). To fourth order in the field components and first order in g, (5.149) yields an energy

$$U' = f_x'^4 u^{(401)} + f_z'^4 u^{(041)} + f_x'^2 f_z'^2 u^{(221)} \tag{5.151}$$

(since the energy U' of this auxiliary problem must be invariant under $f_\alpha' \to -f_\alpha'$). Then inserting the identities (5.150) into (5.151) gives

$$\begin{aligned} U_{\hat{\mathbf{R}}}^{(401)} &= (1-R_x^2)^2 u^{(401)} + R_x^4 u^{(041)} + R_x^2(1-R_x^2) u^{(221)}, \\ U_{\hat{\mathbf{R}}}^{(041)} &= (1-R_z^2)^2 u^{(401)} + R_z^4 u^{(041)} + R_z^2(1-R_z^2) u^{(221)}, \\ U_{\hat{\mathbf{R}}}^{(221)} &= 2(R_y^2 + 3R_x^2 R_z^2) u^{(401)} + 6R_x^2 R_z^2 u^{(041)} \\ &\quad + (R_x^2 + R_z^2 - 6R_x^2 R_z^2) u^{(221)}. \end{aligned} \tag{5.152}$$

These are combined in (5.147) to get the desired energy terms for the reconstructed cubic perturbation.

A simplification arises from noting that in the sum (5.147) the choice of cubic vectors $\hat{\mathbf{R}}$ is arbitrary. For the particular case $l = 4$, choosing first the set of six vectors of the form $(\pm 1, 0, 0)$ and then the set of twelve vectors of the form $\frac{1}{\sqrt{2}}(\pm 1, \pm 1, 0)$, and evaluating (5.147) both ways, yields the identity

$$8u^{(401)} + 3u^{(041)} + 2u^{(221)} = 0 \text{ (for } l = 4). \tag{5.153}$$

If this is substituted back into (5.152) and the sum (5.147) carried out in general, one readily discovers (for $l = 4$)

$$E^{(401)} = E^{(041)} = -\frac{1}{3}E^{(221)} = \sqrt{\frac{3}{7\pi}}\left(2u^{(401)} + u^{(041)}\right). \tag{5.154}$$

This can be further simplified. Expanding first in g and then in powers of f_α, the auxiliary energy terms can be written

$$u^{(\alpha\beta 1)} = \int d^3 r v_{cl}(r) P_l(\hat{\mathbf{r}} \cdot \hat{\mathbf{z}}) n^{(\alpha\beta 0)}. \tag{5.155}$$

Here $n^{(400)}$ and $n^{(040)}$ are the densities induced by electric fields oriented, respectively, along x and z, in the absence of the cubic perturbation:

$$\begin{aligned} n^{(040)} &= n_0(r) + n_2(r)P_2(\hat{\mathbf{r}} \cdot \hat{\mathbf{z}}) + n_4(r)P_4(\hat{\mathbf{r}} \cdot \hat{\mathbf{z}}) \\ n^{(400)} &= n_0(r) + n_2(r)P_2(\hat{\mathbf{r}} \cdot \hat{\mathbf{x}}) + n_4(r)P_4(\hat{\mathbf{r}} \cdot \hat{\mathbf{x}}). \end{aligned} \tag{5.156}$$

Expanding $P_4(\hat{\mathbf{r}} \cdot \hat{\mathbf{x}})$ using the addition theorem, and inserting (5.156) into (5.155) gives, for $l = 4$:

$$u^{(401)} = \frac{3}{8} u^{(041)}. \tag{5.157}$$

Note also that (5.155) and (5.156) explain why only $l = 4$ terms in the cubic potential contribute to γ: for $l > 4$, an expression like (5.153) can be obtained relating $u^{(221)}$ to $u^{(401)}$ and $u^{(041)}$; but from (5.155) and (5.156) the latter energies vanish for $l > 4$.

Summary of Calculation

The following summarizes the method of calculation used by Johnson and Subbaswamy (1989). Given the triple perturbation (5.146), the energies $E^{(041)}$ and $E^{(221)}$ which yield the first-order (in g) contributions to the hyperpolarizabilities γ_{1111} and γ_{1122} via (5.136), (5.137) are needed. To calculate these, consider the much simpler problem which consists of a double perturbation of Legendre polynomials oriented along the z-axis:

$$\delta V = fz + gV_{c4}(r)P_4(\hat{\mathbf{r}} \cdot \hat{\mathbf{z}}) \tag{5.158}$$

and find $W^{(41)}$, the energy term of order $f^4 g$ (called $u^{(041)}$ above). Then

$$E^{(041)} = -\frac{1}{3}E^{(221)} = \frac{1}{4}\sqrt{\frac{21}{\pi}}W^{(41)}, \tag{5.159}$$

and, from (5.136), (5.137), the contributions of the cubic potential to the hyperpolarizabilities are, to first order,

$$\gamma^{(1)}_{1111} = -24E^{(041)} \tag{5.160}$$

$$\gamma^{(1)}_{1122} = -\frac{1}{2}\gamma^{(1)}_{1111}. \tag{5.161}$$

Hence the first-order cubic contribution to the dipole hyperpolarizability has a simple relationship between its diagonal and off-diagonal components. Equally useful from a practical viewpoint is the fact that the detailed calculation need only be carried out for a double perturbation with cylindrical symmetry, which saves a great deal of labor.

Combining the zeroth-order (spherical) and first-order cubic results, we find total electronic dipole hyperpolarizabilities

$$\begin{aligned}\gamma_{1111} &= \gamma^{(0)}_{1111} + \gamma^{(1)}_{1111} \\ \gamma_{1122} &= \frac{1}{3}\gamma^{(0)}_{1111} - \frac{1}{2}\gamma^{(1)}_{1111}.\end{aligned} \tag{5.162}$$

The relative anisotropy $1 - 3\gamma_{1122}/\gamma_{1111}$ will deviate from its spherical value of zero as the cubic perturbation grows.

Most of the computational effort lies in solving for the energy $W^{(41)}$ arising from the perturbation (5.158). The procedure is very similar to that used to find the diagonal component (Johnson *et al.*, 1987). A hierarchy of Kohn-Sham equations is solved self-consistently to find the perturbed wave functions to orders (0,0), (1,0), (0,1), (2,0), and (1,1) in the double perturbation (5.158). Details of how to solve the first four of these have been discussed earlier; the angular momentum decomposition of the latter is entirely similar. It turns out that the energy $W^{(41)}$ can be obtained from these first- and second-order wave functions. A procedure like that outlined in this section can also be used to find simple relations between the diagonal and off-diagonal components of the quadrupole hyperpolarizability B. Adding a field-derivative term to (5.146) and expanding the cubic perturbation as above, it turns out that the contributions to B (to first-order in g) are (see Johnson and Subbaswamy, 1989)

$$B^{(1)}_{11,11} = -2B^{(1)}_{11,22} = -\frac{3}{2}\sqrt{\frac{21}{\pi}}u^{(0211)} \tag{5.163}$$

where $u^{(0211)}$ is the energy term second-order in f_z and first-order in g and the field-derivative term. Finally, this method could be modified for any noncubic crystal by replacing kubic harmonics by the sums of spherical harmonics which transform symmetrically under the relevant point group, and generalizing (5.141).

5.6.3 Diagonal Component

The results for γ in the spherical solid model from the work of Johnson *et al.* (1987) have already been given in Table 5.12. Johnson and Subbaswamy

Table 5.13: Dipole hyperpolarizability for the alkali halides (10^{-39} esu). The zeroth-order results are for the spherical solid model; the first-order cubic corrections are also shown. From Johnson and Subbaswamy (1989).

	Zeroth-order		First-order		Total
	γ^{+}_{1111}	γ^{-}_{1111}	γ^{+}_{1111}	γ^{-}_{1111}	γ_{1111}
LiF	0.174	244.	-0.00	-6.53	237.
NaF	3.61	518.	-0.04	-11.7	510.
KF	62.5	778.	-1.39	-27.8	812.
RbF	174.	1020.	-4.86	-34.7	1150.
LiCl	0.175	1210.	-0.00	-23.9	1180.
NaCl	3.56	2030.	-0.02	-25.3	2010.
KCl	62.2	2750.	-0.92	-80.9	2730.
RbCl	174.	3450.	-3.46	-97.0	3520.
LiBr	0.175	2030.	-0.00	-29.3	2000.
NaBr	3.56	3230.	-0.02	-24.2	3210.
KBr	61.7	4190.	-0.79	-111.	4140.
RbBr	172.	5220.	-2.99	-132.	5260.
LiI	0.176	3780.	-0.00	-22.6	3760.
NaI	3.56	5790.	-0.01	2.92	5800.
KI	61.3	7280.	-0.62	-147.	7200.
RbI	170.	8830.	-2.42	-176.	8820.

(1989) calculated the cubic correction to γ_{1111} by the method described above. Their results are shown in Table 5.13. The total diagonal element γ_{1111} is changed only slightly from its isotropic value $\gamma^{(0)}_{1111}$, calculated in the spherical solid model and reported in Section 5.6.1.

Before comparing the calculated values with experiments, some comments concerning the experimental results are in order. The techniques used are (see, e.g., Hellwarth, 1977) (1) third harmonic generation (THG), which measures $\chi^{(3)}(3\omega; -\omega, -\omega, -\omega)$; (2) four-wave mixing (FWM), which measures $\chi^{(3)}(\omega_2 - 2\omega_1; \omega_1, \omega_1, -\omega_2)$; and (3) nonlinear refractive index (NRI) measurements, which give $\chi^{(3)}(-\omega; \omega, \omega, -\omega)$. The most recent, and most accurate results are those of Adair *et al.* (1989) from NRI measurements. These authors have also reviewed the previous experimental results for the alkali halides. We will use their results for comparison with experiments.

The relationship between the macroscopic quantity $\chi^{(3)}$ and the microscopic quantity γ has been given in Section 4.4. The experimental numbers need to be extrapolated to the static values $\chi^{(3)}(0;0,0,0)$ for comparison with calculated results. For this purpose Johnson *et al.* (1987) used the semi-empirical formula (see Elliott and Ward, 1984; Mizrahi and Shelton, 1985)

$$\chi^{(3)}(\omega_1;\omega_2,\omega_3),\omega_4 = \chi^{(3)}(0,0,0)(1+c\omega_L^2/\omega_0^2), \tag{5.164}$$

where ω_0 is determined from the dispersion of the linear susceptibility,

$$\chi^{(1)}(\omega) = (1+\omega^2/\omega_0^2), \tag{5.165}$$

and

$$\omega_L^2 = \sum_{i=1}^{4} \omega_i^2, \tag{5.166}$$

and c is a number of order 10. The work on rare gases (Mizrahi and Shelton, 1985) suggests that we take $c \approx 3$. To determine ω_0 Johnson *et al.* used the refractive index dispersion data of Lowndes and Martin (1969). It is interesting to note that the ω_0 obtained in this way scale reasonably well with the first optical absorption peaks (Woodruff, 1983) in these crystals. From the extrapolated static values of $\chi^{(3)}$ the experimental values for the static hyperpolarizabilities γ for the alkali may be determined from (4.98). The calculated $\gamma^+ + \gamma_-$ are compared with the γ deduced from the NRI measurements of Adair *et al.* in Table 5.14. The agreement is reasonable, considering the known accuracy of both the experimental and theoretical values.

The only other first-principles theoretical calculation is that of Fowler and Madden (1985) for LiF and LiCl by quantum chemical methods. For LiF they find $\gamma_{1111} = 105 \times 10^{-39}$ esu, while for LiCl they find $\gamma_{1111} = 1050 \times 10^{-39}$ esu at zero frequency; the TDLDA results are 237×10^{-39} esu, and 1180×10^{-39}esu, respectively. As seen throughout this book, the Hartree-Fock based results are consistently lower than the TDLDA results.

Adair *et al.* (1989) attempted to extend the Wilson-Curtis empirical model of anion linear polarizability, discussed in Section 5.2, to the dipole hyperpolarizability as well. They proposed writing

$$\gamma^- = \gamma_0^- \exp(-D/R^2), \tag{5.167}$$

Table 5.14: Dipole hyperpolarizability for the alkali halides (10^{-39} esu). The total calculated hyperpolarizabilities shown are the sum of the cation and anion contributions. From Johnson and Subbaswamy (1989).

	γ_{1111}(total)	
	TDLDA[a]	Expt.[b]
LiF	252	141
NaF	549	320
KF	874	889
NaCl	2340	1630
KCl	3170	3050
NaBr	3970	3370
KBr	5040	4930

[a] Extrapolated to the experimental frequency
[b] Adair *et al.* (1989)

where, as before (see (5.30)), R is the nearest-neighbor distance in the crystal, and D is a constant for a particular anion. A simple scaling argument,

$$\begin{aligned} \alpha &\propto \; < r^2 >, \\ \gamma &\propto \; < r^4 >, \end{aligned}$$

would suggest $D = 2C$, where C is the constant that enters into the Wilson-Curtis expression for α^-. They found that the calculated γ^- values of Johnson *et al.* (1987) fit the form (5.167) quite well, with the values of γ_0^- and D shown in Table 5.15. The simple expectation $D = 2C$ is not borne out, indicating that the simple scaling argument does not work. Nonetheless, the formula (5.167), with the parameters given in Table 5.15, provides a useful representation of the hyperpolarizabilities of alkali halide crystals.

5.6.4 Anisotropy Ratio

Measurements of the ratios between the various independent components of the nonlinear susceptibility tensor are more accurate than the difficult task of measuring absolute quantities; this permits a more demanding test

Table 5.15: Parameters in the Wilson-Curtis formula for anion hyperpolarizability. From Adair *et al.* (1989).

	γ_0^- (10^{-39}esu)	Slope D (Å^2)
F^-	4040	11.3
Cl^-	16100	16.9
Br^-	24400	18.6
I^-	47500	22.6

of the various theoretical contributions to the susceptibility. Moreover, the off-diagonal components themselves are important in the calculation of important physical processes, such as Raman scattering intensity. Johnson and Subbaswamy (1989) performed a TDLDA calculation of the off-diagonal component γ_{1122} for the alkali halides, along lines described above. Their results are presented below.

A number of physical processes jointly give rise to $\chi^{(3)}$. For our purposes, they can be grouped into two categories, the electro-optic contribution (electronic response) and the elasto-optic contribution (nuclear or ionic motion). The electrons respond very quickly and can often be taken to adiabatically follow the electric field and the nuclear motion. This, the Born-Oppenheimer approximation, is valid for nonlinear effects at optical frequencies (i.e., for processes in which all frequencies are well above the nuclear-response frequencies but far below electronic frequencies). The consequences of the Born-Oppenheimer approximation for third-order nonlinear optical processes were developed in detail by Hellwarth (1977), from which we have extracted the results useful here. In the case of the nonlinear refractive index (NRI) measurement, which is of particular interest here, the Born-Oppenheimer third-order susceptibility is the tensor

$$\chi^{(3)}_{ijkl}(-\omega;\omega,\omega,-\omega) = \chi^{(3)el}_{ijkl} + \frac{1}{6}D_{ijkl}(0), \tag{5.168}$$

where $\chi^{(3)}_{el}$ is a constant tensor describing the third-order electronic response, and D_{ijkl} represents the contribution of nuclear motion. An experimentally measured third-order effect combines the electronic and nuclear contributions. However, D_{ijkl} is directly related to the cross section for Raman scattering (Hellwarth, 1977) and so can be obtained independently in principle, permitting one to separate the contributions to (5.168).

In a cubic system, $\chi^{(3)el}$ has only two independent components and D has three. The tensors can be written

$$\chi^{(3)el}_{ijkl} = \left(\frac{1}{3}\Gamma_0 - \frac{1}{2}\Gamma_1\right)(\delta_{ij}\delta_{kl} + \delta_{ik}\delta_{jl} + \delta_{il}\delta_{jk}) + \frac{5}{2}\Gamma_1\delta_{ijkl}, \tag{5.169}$$

$$\begin{aligned} D_{ijkl}(\omega) &= [A(\omega) - 2E(\omega)]\delta_{ij}\delta_{kl} + T(\omega)(\delta_{ik}\delta_{jl} + \delta_{il}\delta_{jk}) \\ &\quad + [6E(\omega) - 2T(\omega)]\delta_{ijkl}, \end{aligned} \tag{5.170}$$

where $\delta_{ijkl} = 1$ if $i = j = k = l$ and 0 otherwise. The form of (5.170) has been chosen to correspond to the usual Raman notation (A, E, and T referring to the various symmetry components of the spectra) (see, e.g., Hayes and Loudon, 1978), and (5.169) is written in a form convenient for what follows. In an isotropic medium (5.169) and (5.170) hold with $\Gamma_1 = 0$ and $T(\omega) = 3E(\omega)$.

The nonlinear optical response of a cubic system, unlike its linear response, is not isotropic. A linearly polarized beam propagating along one crystal axis and polarized at an angle θ relative to another has an effective third order susceptibility (Adair *et al.*, 1989)

$$\chi^{(3)}_{\text{eff}}(\theta) = \chi^{(3)}_{1111} - \left(\chi^{(3)}_{1111} - 2\chi^{(3)}_{1122} - \chi^{(3)}_{1221}\right)\frac{\sin^2 2\theta}{2}. \tag{5.171}$$

When Kleinman symmetry $[\chi^{(3)}_{1122} = \chi^{(3)}_{1212}]$ is valid (which is true in the Born-Oppenheimer approximation if the electronic contribution dominates), (5.171) becomes

$$\chi^{(3)}_{\text{eff}}(\theta) = \chi^{(3)}_{1111} - (\chi^{(3)}_{1111} - 3\chi^{(3)}_{1122})\frac{\sin^2 2\theta}{2}. \tag{5.172}$$

The second piece on the right-hand side of (5.172) represents the optical anisotropy occurring in third order. We thus define the *cubic anisotropy* $\Delta\chi^{(3)}$ of the third-order susceptibility as

$$\Delta\chi^{(3)} = \chi^{(3)}_{1111} - 3\chi^{(3)}_{1122}. \tag{5.173}$$

Similarly, the anisotropy of the hyperpolarizability is defined as

$$\Delta\gamma = \gamma_{1111} - 3\gamma_{1122}, \tag{5.174}$$

which vanishes only in the case of a spherically symmetric environment. Thus $\Delta\chi^{(3)}$ or $\Delta\gamma$ is a measure of the optical anisotropy caused in third order by the environment's nonsphericity. In the limiting case where the susceptibility is dominated by the purely electronic contribution, the fractional anisotropies are equal: $\Delta\chi^{(3)}/\chi^{(3)}_{1111} = \Delta\gamma/\gamma_{1111}$.

Table 5.16: Off-diagonal dipole hyperpolarizability for the alkali halides (10^{-39} esu). The sum of the zeroth-order contribution ($= \frac{1}{3}\gamma_{1111}^{(0)}$) and the first-order contribution ($\gamma_{1122}^{(1)}$) is shown. From Johnson and Subbaswamy (1989).

	γ_{1122}		γ_{1122}
LiF	85	LiBr	693
NaF	180	NaBr	1090
KF	295	KBr	1470
RbF	416	RbBr	1870
LiCl	414	LiI	1270
NaCl	692	NaI	1930
KCl	980	KI	2520
RbCl	1260	RbI	3090

Results for Alkali Halides

Johnson and Subbaswamy (1989) carried out the calculation of $\gamma_{1122}^{(1)}$ for alkali halide crystals. The crystal environment was described by a sum of external potentials V_R centered on each ionic site (see (5.142)); for the nearest four ionic shells pseudopotentials were used, while more distant ions were treated as point charges. The pseudopotentials are those given in (5.37) and (5.38). The results of Johnson and Subbaswamy (1989) are shown in Table 5.16. The sum of the zeroth-order contribution ($= \frac{1}{3}\gamma_{1111}^{(0)}$) and the first-order contribution ($\gamma_{1122}^{(1)}$) is shown.

In comparing various contributions to the anisotropy, it is more convenient to use the macroscopic third-order nonlinear susceptibility $\chi^{(3)}$ rather than the microscopic hyperpolarizability γ. Comparing the general cubic expressions for $\chi^{(3)el}$, (5.169), and γ, (5.162), shows that $\Gamma_0 \propto \gamma_{1111}^{(0)}$ and $\Gamma_1 \propto \gamma_{1111}^{(1)}$. Then the electronic contribution to the susceptibility anisotropy (5.173) is, using (4.98),

$$\Delta\chi_{el}^{(3)} = \frac{5}{2}\Gamma_1 = \frac{5f^4}{48V_0}\gamma_{1111}^{(1)}. \tag{5.175}$$

Table 5.17 shows the electronic (from LDA theory) and experimental susceptibility anisotropies $\Delta\chi^{(3)}$ for several of the alkali halide crystals. The measured values (Adair *et al.*, 1989) have been extrapolated to zero frequency

Table 5.17: Cubic anisotropy $\Delta\chi^{(3)}$ of alkali halide crystals. From Johnson and Subbaswamy (1989).

	TDLDA	Expt.[a]
LiF	-0.12	-0.34±0.05
NaF	-0.12	-0.38±0.04
KF	-0.22	+0.31±0.18
NaCl	-0.24	-0.66±0.21
KCl	-0.49	+0.62±0.27
NaBr	-0.20	-3.1±2.2
KBr	-0.71	-3.4±3.8

[a] Adair *et al.* (1989)

for consistency as described above. Relative anisotropies would be dispersionless in the simplest model. Despite the fairly large uncertainty in the measured anisotropy, it is evident that the theoretical electronic anisotropies are smaller (35% or less) than the measured values. Notice that in two cases (KF and KCl), the observed anisotropies differ in sign from the calculated electronic contributions.

The elasto-optic or lattice contribution to the susceptibility is relatively small; ordinarily the nearly-instantaneous electronic response dominates at optical frequencies. However, the cubic anisotropy (5.173) of the third-order susceptibility can, for ions in nearly isotropic environments, be much smaller than $\chi^{(3)}_{1111}$. *A priori* a substantial lattice contribution to $\Delta\chi^{(3)}$ cannot be ruled out.

We can get an estimate of the lattice contribution to $\chi^{(3)}$ in the Born-Oppenheimer approximation following Hellwarth (1977). For definiteness, consider a nonlinear refractive index experiment; the susceptibility $\chi^{(3)}$ is given in the Born-Oppenheimer approximation in (5.168). Using (5.169) and (5.170), the Born-Oppenheimer expression for the cubic anisotropy (5.173) is

$$\Delta\chi^{(3)} = \Delta\chi^{(3)}_{el} - \frac{1}{3}(A(0) - 5E(0)). \tag{5.176}$$

$\Delta\chi^{(3)}_{el}$ is the electronic contribution (5.175); the remainder is the contribution from nuclear or ionic motion. The imaginary part of $D(\omega)$ is proportional to the cross section for Raman scattering (second-order scattering in the case of cubic crystals). From $\operatorname{Im} D$, the real part of D can be obtained via a Kramers-Kronig relation; then $D(0)$ gives the lattice contribution to the susceptibility (5.168).

This approach requires an accurate measurement of the *absolute* second-order Raman intensity. Unfortunately, measurements of absolute cross sections are scarce, and often not very reliable. We estimate the lattice contribution for one example, KCl, using a published Raman spectrum (Krantz and Luty, 1985). Approximating the intensity by a straight line cut off at 300 cm^{-1}, we estimate $\mathrm{Re}\, D_{1111}(0) = 4.1 \times 10^{-17}$ esu for KCl. Scattering experiments at different polarizations for various alkali halides (Krauzman, 1967; 1968) show $E(0)$ and $T(0)$ to be an order of magnitude smaller than $A(0)$; hence the lattice contribution to the anisotropy for KCl is

$$\Delta\chi^{(3)}_{lattice} \approx -\frac{1}{3}A(0) \approx -\frac{1}{3}D_{1111}(0) \approx -1.4 \times 10^{-18}\text{esu} \approx -0.002\chi^{exp}_{1111} \tag{5.177}$$

where χ^{exp}_{1111} is the measured diagonal susceptibility (Adair *et al.*, 1989) (extrapolated to $\omega = 0$ as described above). The electronic contribution to the anisotropy is, from Table 5.17, $-0.07\chi^{exp}_{1111}$. Evidently the contribution to the anisotropy from nuclear motion is smaller than the electronic anisotropy by an order of magnitude.

Both the diagonal and off-diagonal components of the hyperpolarizability are dominated by the anion's contribution (see Table 5.13). As a given anion is paired with different cations in the alkali halides, γ_{ijkl} varies substantially. The variation in the diagonal hyperpolarizability can be explained by changes in the cationic compression of the anion's wave function as discussed in Section 5.6.1 One might then reasonably conclude that the anisotropy in the hyperpolarizability could be similarly explained, since the compression would be greatest along [100] (Adair *et al.*, 1989). The present results suggest that this is not the case–the purely electronic contribution is unable to explain the variation in the measured anisotropy $\Delta\chi^{(3)}$ (Table 5.17).

One other possible source of aniostopy in the nonlinear susceptibility is an anisotropic local field correction. The hyperpolarizability computed by Johnson and Subbaswamy (1989) results from the intrinsic nonlinear response of a single ion's electrons. An ion in isolation would possess such a nonlinear reponse. Johnson and Subbaswamy represented the crystalline environment by placing the ion in a cage of unpolarized ions (pseudopotentials). The effect of polarizing all other ions in the medium is incorporated into the theory by the usual local field corrections in going from the hyperpolarizability to the susceptibility (see (4.98)). This amounts to an isotropic rescaling of the field. Both the crystalline cage and the local field corrections modify the nonlinear response of the ion, but they do not themselves produce any nonlinearity. The cubic cage results in the optical anisotropy, whereas the local field corrections in this approximation do not. However, Akhmanov *et al.* (1977) have developed a microscopic model of the local field, based on the shell model, in which the local field itself can produce nonlinear, optical response.

In the shell model, with fixed nuclei, the ion at site $\vec{R}$ is replaced by an oscillator of charge $e_{\vec{R}}$ (representing the outer electronic shell) bound to the ionic core. A shell at, say, $\vec{R} = 0$ is subject to a local electric field made up of the external field $\vec{\mathcal{E}}_{\text{ext}}$, the field $\vec{\mathcal{E}}'$ from polarized ions within the Lorentz cavity, and $\vec{\mathcal{E}}''$ from ions outside the cavity. Let $\vec{w}_R$ be the deviation of the shell at $\vec{R}$ from equilibrium; then the field $\vec{\mathcal{E}}'$ at $\vec{R} = 0$ is

$$\mathcal{E}'(0) = \nabla_R \sum_{\vec{R}\neq 0} \frac{e_{\vec{R}}}{|\vec{R} + \vec{w}_R|}. \tag{5.178}$$

If all ions of the same type ($\alpha = a$ or c, meaning anions or cations) undergo the same small deviation w_α, this can be expanded for a cubic crystal:

$$\mathcal{E}_i'(0) = \sum_{\alpha=a,c} \left(f^{(1)}_{\alpha,ij} w_{\alpha j} + f^{(3)}_{\alpha,ijkl} w_{\alpha j} w_{\alpha k} w_{\alpha l} + \cdots \right) \tag{5.179}$$

with

$$f^{(n)}_{\alpha,ijk\cdots n} = \frac{1}{N!} \sum_{\vec{R}\neq 0}^{(\alpha)} \frac{\partial^{n+1}}{\partial R_i \partial R_j \cdots \partial R_n} \frac{1}{R} \tag{5.180}$$

where the sum in (5.180) is over the type-α sites within the Lorentz cavity. The first point to note here is that the local field itself possesses nonlinearity and contributes to the third-order susceptibility in this model. In perturbation theory, replacing $w_{\alpha j}$ in the second term in (5.179) by its lowest-order value ($\propto \mathcal{E}_j^{ext}$) gives rise to a third-order contribution to the local field. This produces an additional contribution to the hyperpolarizability not included in our calculation, even if the individual shell oscillators are harmonic, coupling them via the Coulomb interaction produces a third-order response.

Moreover, this additional third-order response is anisotropic. One can readily show from (5.180) that in a cubic system $f^{(3)}_{\alpha,1122} = -\frac{1}{2} f^{(3)}_{\alpha,1111}$, yielding contributions to the susceptibility which also satisfy $\chi^{(3)}_{1122} = -\frac{1}{2}\chi^{(3)}_{1111}$. This is the same relation we found for the TDLDA calculation. To show clearly the origin of this relationship we consider a simple driven oscillator model

(Johnson and Subbaswamy, 1989) on which most phenomenological models are based.

Suppose that the oscillator (of charge e) has a small anharmonic potential with cubic symmetry; in particular, let the potential go as the $l = 4$ symmetric kubic harmonic $\alpha_4(\hat{\mathbf{r}})$. A simple example is $V_{\text{anh}} = m\alpha(x^4 + y^4 + z^4 - \frac{3}{5}r^4)/4$. In the presence of a static electric field $\mathbf{E}$, the oscillator's equation of motion is

$$\ddot{\vec{r}} + \omega^2\vec{r} + \alpha(x^3\hat{\mathbf{i}} + y^3\hat{\mathbf{j}} + z^3\hat{\mathbf{k}} - \frac{3}{5}r^3\hat{\mathbf{r}}) = \frac{e}{m}\vec{\mathcal{E}}. \tag{5.181}$$

To first order in α, the resulting equilibrium dipole moment is $\mathbf{p}$, where

$$p_x = ex = \frac{e^2\mathcal{E}_x}{m\omega^2} - \frac{\alpha e^4}{5m^3\omega^8}(2\mathcal{E}_x^3 - 3\mathcal{E}_x\mathcal{E}_y^2 - 3\mathcal{E}_x\mathcal{E}_z^2) \tag{5.182}$$

with similar expressions for p_y and p_z. This gives

$$\begin{aligned}\gamma_{1111} &= -\frac{12\alpha e^4}{5m^3\omega^8} \\ \gamma_{1122} &= \gamma_{1111}\cdot\left(-\frac{3}{2}\right)\cdot\frac{1}{3} = -\frac{1}{2}\gamma_{1111}.\end{aligned} \tag{5.183}$$

The interesting result here is that an $l = 4$ (i.e., α_4) cubic anharmonicity in the potential gives contributions to γ related by the simple property $\gamma_{1122} = -\frac{1}{2}\gamma_{1111}$. Recall that this relationship was also realized in the evaluation of the contribution to first order in the cubic potential in the TDLDA calculation described above (see (5.161)). In the case of spherical symmetry, the off-diagonal piece is related to the diagonal by $\gamma_{1122} = \frac{1}{3}\gamma_{1111}$, as would result above if, for example, one used a spherically symmetric form such as $\frac{1}{4}\alpha r^4$ for the anharmonic potential V_{anh}.

It is difficult to estimate the magnitude of this local field contribution to the anisotropy of the susceptibility. Akhmanov *et al.* (1977) ascribed the entire nonlinear response to these local field nonlinearities, neglecting the intrinsic nonlinearity of the ions. (This can be thought of in their language as an isotropic anharmonicity in the spring binding a shell to its core.) This is a serious omission. We have seen that the intrinsic nonlinearity explains the diagonal hyperpolarizabilities of the alkali halides rather well; this implies that the spherically symmetric contribution to γ_{1122} is substantial ($\gamma^{(0)}$ in (5.162)). Moreover, their neglect of the intrinsic nonlinearity led Akhmanov *et al.* to conclude that the relative anisotropy should be exactly

$\Delta\gamma/\gamma = 5/2$ for the alkali halides. Nevertheless, using further approximations they obtained diagonal hyperpolarizabilities for the alkali halides of the order of those measured. This may in fact indicate a qualitative truth–that the nonlinearities in the local field can contribute significantly to the hyperpolarizability and to the anisotropy as well. More work is needed to resolve the disagreement in the anisotropy ratio between theory and experiment.

5.7 Raman Scattering Intensity

To highlight the importance of the calculation of the nonlinear polarizabilities of the type discussed in the previous section we consider Raman scattering in alkali halide crystals in this section. We follow closely the discussion in Mahan and Subbaswamy (1986). Until fairly recently, Raman scattering studies have been confined to line shape, polarization, and relative intensity measurements. Accordingly, theoretical analyses have been content with treating the relevant matrix elements ("Raman tensor") as phenomenological parameters subject only to group theoretical symmetry requirements. However, in recent years measurement of absolute intensities is becoming routine, and there is interest in calculating the absolute values of the relevant parameters.

First-order Raman scattering is not allowed in centrosymmetric crystals (see, e.g., Hayes and Loudon, 1978). However, second-order scattering, whereby the photon is scattered by the creation of two phonons of wave vector $\vec{q}$ and $-\vec{q}$, is allowed and observed experimentally. The resulting Raman spectrum bears a strong resemblance to the joint density of two-phonon states.

A general quantum mechanical derivation of the Raman tensor in crystals was given by Loudon (1963). The practical evaluation using this approach is hindered by the need for reliable Bloch wave functions for electrons in excited states in the crystal. The traditional approach, originally due to Born and Bradburn (1946), is to expand the crystal polarizability in powers of the phonon amplitude, with the relevant polarizability derivatives treated as disposable parameters (see, e.g., Hardy and Karo, 1979). This involves as many as ten parameters in additon to those used in the lattice dynamics calculation in order to get satisfactory fit to experimental line shapes. In view of the several competing contributions, namely, electron overlap, dipole-induced dipole, electro-optic, the parameters cast no light on the relative importance of these effects.

Mahan and Subbaswamy suggested a different approach, taking advantage of the closed-shell electronic configuration of the ions, and the fact that the probe energy (~ 2.4 eV) is much smaller than the lowest electronic resonance (≥ 6 eV). They proposed starting with a localized description traditionally used for molecules and liquids (Berne and Pecora, 1976). The polarization density of the medium is written as

$$\vec{P}(\vec{r},t) = \sum_i \vec{p}^{(i)}(\{R\};t)\delta(\vec{r} - \vec{R}_i(t)), \tag{5.184}$$

where the instantaneous dipole moment of the ith entity (atom, ion, or molecule), which depends on the position of all others, is given by

$$\begin{aligned} p_\alpha^{(i)}(\{\vec{R}\};t) &= \alpha_{\alpha\beta}^{(i)}(\{\vec{R}\};t)\bar{\mathcal{E}}_\beta(\vec{R}_i,t) \\ &+\frac{1}{6}\gamma_{\alpha\beta\mu\nu}^{(i)}\bar{\mathcal{E}}_\beta\bar{\mathcal{E}}_\mu\bar{\mathcal{E}}_\nu \\ &+\frac{1}{3}B_{\alpha\beta,\mu\nu}^{(i)}\bar{\mathcal{E}}_\beta\bar{Q}_{\mu\nu} + \cdots . \end{aligned} \tag{5.185}$$

In the last two terms on the right hand side we have suppressed the arguments of the various quantities for brevity. $\bar{\mathcal{E}}_\alpha$ and $\bar{Q}_{\mu\nu}$ are, respectively, the local field and field gradient components. The fact that the ions are spherical has been used to eliminate some terms.

In (5.184) there are three separate contributions to fluctuations in the polarization: First there is the density factor, which by itself gives rise to Brillouin scattering. The various order polarizabilities, α, γ, B, etc., depend implicitly on the instantaneous positions of all other ions due to electron overlap. If fixed values are assigned to these quantities, neglecting overlap effects, then the fluctuations in $\vec{P}$, which give rise to inelastic light scattering, arise only from fluctuations in the local field. We term this *the generalized electro-optic coupling.* For large interionic separations this is the only relevant effect. For small separations the polarizabilities are themselves modulated due to nuclear displacements, and this gives rise to what we term the *atomic displacement coupling.* In polar semiconductors, for the long-wavelength optic phonon, the macroscopic electric field and the atomic displacement are proportional to each other via frequency dependent factors. This gives rise to well-known interference effects in the Raman spectrum of polaritons in these materials. For the alkali halides, where phonons from the entire zone contribute, such a general relation can not be obtained, and the two contributions should be calculated separately. The usual Born-Bradbury type of analysis, where the polarizability derivative parameters are frequency-independent quantities, amounts to considering only the atomic displacement part of the Raman tensor. Mahan and Subbaswamy considered the electro-optic contribution in detail.

5.7.1 Local Field

In Section 4.4 an expression for the local electric field in terms of the applied field was given for a uniform applied field. In the case of fields due to ionic displacements, especially those involving large wave vector phonons, there are large gradients to deal with. Mahan and Subbaswamy (1986) considered this question in detail, and their results are summarized here.

Local field corrections arise because the electric field at an ion site is different from the average value of the field within the unit cell. These differences are caused by the atomic nature of the solid. Local fields cause the ions to polarize. The polarization creates fields which contribute to the field at other sites. The mutual interaction of the fields among the ions must be solved self-consistently. The method for deriving the local fields used by Mahan and Subbaswamy (1986) is generalization of the method given by Mahan (1967) for linear media. Let $p_\alpha^{(j)}$ and $q_{\alpha\beta}^{(j)}$ be the induced dipole and quadrupole moments on the ion j, respectively. Define a Hamiltonian which describes the interaction of the induced moments among themselves, as well as the interaction of these induced moments with the external electric field $\mathcal{E}$ and quadrupole field Q:

$$
\begin{aligned}
H &= -p_\alpha^{(j)}\mathcal{E}_\alpha^{(j)} - q_{\alpha\beta}^{(j)}Q_{\alpha\beta}^{(j)} \\
&\quad -\frac{1}{2}\left(p_\alpha^{(j)}\phi_{\alpha\beta}^{(jk)}p_\beta^{(k)} + 2p_\alpha^{(j)}\Psi_{\alpha\beta\gamma}^{(jk)}q_{\beta\gamma}^{(k)} + \sum_\delta q_{\alpha\beta}^{(j)}\Xi_{\alpha\beta\gamma\delta}^{(jk)}q_{\gamma\delta}^{(k)}\right),
\end{aligned}
\tag{5.186}
$$

where summation over repeated indices and superscripts is implied. Only dipole and quadrupole terms are included since octupole contributions are expected to be small. The quantities Φ, Ψ, Ξ are the multipole series for the Coulomb interaction:

$$
\begin{aligned}
\Phi_{\mu\nu} &= \nabla_\mu\nabla_\nu(1/r), \\
\Psi_{\mu\nu\lambda} &= \nabla_\mu\nabla_\nu\nabla_\lambda(1/r), \\
\Xi_{\alpha\beta\mu\nu} &= \nabla_\alpha\nabla_\beta\nabla_\mu\nabla_\nu(1/r).
\end{aligned}
\tag{5.187}
$$

Greek subscripts denote xyz, while the roman indices denote ion sites. Energy terms are found by expanding the free energy in a linked cluster expansion (Mahan, 1967) using

$$\exp(-\beta\Omega) = Tr[\exp(-\beta H)]. \tag{5.188}$$

Polarizabilities and hyperpolarizabilities are introduced into this expansion as correlation functions of the induced moments:

$$\begin{aligned}
\alpha_{\mu\nu} &= \beta^{-1}\int d\tau_1 \int d\tau_2 < p_\mu(\tau_1)p_\nu(\tau_2) >,\\
\gamma_{\alpha\beta\mu\nu} &= \beta^{-1}\int d\tau_1 \int d\tau_2 \int d\tau_3 \int d\tau_4 < p_\alpha(\tau_1)p_\beta(\tau_2)p_\mu(\tau_3)p_\nu(\tau_4) >,\\
B_{\alpha\beta\mu\nu} &= -\frac{3}{\beta}\int d\tau_1 \int d\tau_2 \int d\tau_3 < p_\alpha(\tau_1)p_\beta(\tau_2)q_{\mu\nu}(\tau_3) >,\\
C_{\alpha\beta\mu\nu} &= \frac{3}{\beta}\int d\tau_1 \int d\tau_2 < q_{\alpha\beta}(\tau_1)q_{\mu\nu}(\tau_2) > .
\end{aligned} \tag{5.189}$$

The τ integrals are over the interval $(0, \beta)$. These definitions are chosen to agree with those of Buckingham (1967). The linear octupole polarizability C is one-third of the usual definition of this quantity. In the expansion of (5.188) terms of order H^n contribute to the free energy Ω_n. By summing all of the separate contributions to Ω, the total interaction energy W may be written, limiting ourselves to terms that contain only one power of γ and B, as

$$\begin{aligned}
W &= -\frac{1}{2}\left(\mathcal{E}_\alpha^{(j)}\alpha_{\alpha\beta}^{(j)}\bar{\mathcal{E}}_\beta^{(j)} + \frac{1}{3}Q_{\alpha\beta}^{(j)}C_{\alpha\beta\mu\nu}^{(j)}\bar{Q}_{\mu\nu}^{(j)} + \frac{1}{3}B_{\alpha\beta\mu\nu}^{(j)}\bar{Q}_{\mu\nu}^{(j)}\bar{\mathcal{E}}_\alpha^{(j)}\bar{\mathcal{E}}_\beta^{(j)}\right)\\
&\quad -\frac{1}{4!}\gamma_{\alpha\beta\mu\nu}^{(j)}\bar{\mathcal{E}}_\alpha^{(j)}\bar{\mathcal{E}}_\beta^{(j)}\bar{\mathcal{E}}_\mu^{(j)}\bar{\mathcal{E}}_\nu^{(j)},
\end{aligned} \tag{5.190}$$

where again summation over repeated indices is implied. Here the local fields obey the self-consistent (Dyson) equations

$$\begin{aligned}
\bar{\mathcal{E}}^{(j)} &= \mathcal{E}^{(j)} - \sum_k \left\{\Phi^{(jk)}\bar{\mathcal{E}}^{(k)} + \frac{1}{3}\Psi^{(jk)}C^{(k)}\bar{Q}^{(k)}\right\}\\
&= \sum_k \{G^{(jk)}\mathcal{E}^{(k)} + D^{(jk)}Q^{(k)}\},\\
\bar{Q}^{(j)} &= Q^{(j)} - \sum_k \left\{\frac{1}{3}\Xi^{(jk)}C^{(k)}\bar{Q}^{(k)} + \Psi^{(jk)}\alpha^{(k)}\bar{\mathcal{E}}^{(k)}\right\}\\
&= \sum_k \{M^{(jk)}Q^{(k)} + N^{(jk)}\mathcal{E}^{(k)}\},
\end{aligned} \tag{5.191}$$

where, in the interest of brevity, we have suppressed the Cartesian indices; we shall do so in the remainder of this section when no confusion arises. The induced dipole moment on an ion is the functional derivative of the energy with respect to the field on that ion,

$$p_\alpha^{(j)} = -\delta W/\delta \mathcal{E}_\alpha^{(j)}. \tag{5.192}$$

By taking the functional derivative, we get

$$\begin{aligned} p^{(j)} &= \alpha^{(j)}\mathcal{E}^{(j)} + \sum_k \left\{ \frac{1}{6} Q^{(k)} C^{(k)} (\delta \bar{Q}^{(k)}/\delta \mathcal{E}^{(j)}) \right. \\ &\quad + \frac{1}{6} B^{(k)} [2\bar{Q}^{(k)} \bar{\mathcal{E}}^{(k)} \delta \bar{\mathcal{E}}^{(k)}/\delta \mathcal{E}^{(j)} + \bar{\mathcal{E}}^{(k)^2} \delta \bar{Q}^{(k)}/\delta \mathcal{E}^{(j)}] \\ &\quad \left. + \frac{1}{3!} \gamma^{(k)} \bar{\mathcal{E}}^{(k)^3} \delta \bar{\mathcal{E}}^{(k)}/\delta \mathcal{E}^{(j)} \right\}. \end{aligned} \tag{5.193}$$

The functional derivatives in (5.193) are just the Green's functions G, D, M, N introduced in (5.191). Thus, finally, we get the following expression for the induced dipole moment of the ith ion,

$$\begin{aligned} p^{(j)} &= \alpha^{(k)} G^{(kj)} \mathcal{E}^{(k)} + \frac{1}{6} N^{(kj)} [C^{(k)} Q^{(k)} + B^{(k)} (G^{(km)} \mathcal{E}^{(m)})^2] \\ &\quad + \frac{1}{6} G_{kj} [\gamma^{(k)} (G^{(km} \mathcal{E}^{(m)})^3 + 2B^{(k)} (M^{(km)} Q^{(m)} (G^{(kn)} \mathcal{E}^{(n)}], \end{aligned} \tag{5.194}$$

where summation over repeated indices is implied. This expression is the basis for the discussion of electro-optic contribution to Raman scattering. It is exact to first order in γ and B, and includes multipole components up to quadrupole. Only the last two terms contribute to Raman scattering in alkali halides.

5.7.2 Field-Induced First-Order Scattering

First-order Raman scattering is not allowed for the alkali halide crystals. The high degree of cubic symmetry prevents the usual scattering process whereby a photon can scatter while creating or destroying a single long-wavelength phonon. When a static dc electric field is applied to the sample, it breaks the cubic symmetry, and permits a degree of first-order Raman scattering proportional to the square of the applied static field. The formal theory of the effect was written down a long time ago (Burstein, 1966), but the experimental observation was reported much later (Krantz and Lüty, 1985). Dick (1985) gave a theory based on the introduction of nonlinear parameters into the shell model. Subbaswamy and Mahan (1985) based their theory on the electro-optic contribution discussed above.

For field induced first-order Raman scattering (FIFORS) the terms in (5.194) which involve B vanish because only long-wavelength phonons are excited; all the wave vectors in the problem are small. The terms involving B always have the factor of

$$\psi^{(j)}(\vec{q}) = \sum_k \Psi^{(jk)} \exp\left[i\vec{q}\cdot(\vec{r}_j - \vec{r}_k)\right] \tag{5.195}$$

which vanishes for small q because of cubic symmetry. Thus, the γ terms are the only important terms.

For the sake of simplicity, and to emphasize the physics involved, we will first outline the theory neglecting local field effects. The electric field has three sources: the dc field $\mathcal{E}_0$, the laser field $\mathcal{E}_l$, and the field $\mathcal{E}_p$ from the phonon,

$$\mathcal{E} = \mathcal{E}_0 + \mathcal{E}_l + \mathcal{E}_p. \tag{5.196}$$

The quantity $\mathcal{E}^3$ contains 6 times the expression $\mathcal{E}_0\mathcal{E}_l\mathcal{E}_p$. Since γ is independent of the order of its four subscripts, all these combinations are equivalent. As seen in the previous section, the hyperpolarizability of the negative ion is much larger than that of the positive ion, and we consider only the contribution of the anion. Then, the FIFORS cross section can be written as

$$\frac{d\sigma}{d\Omega} = \left(\frac{\omega_f}{c}\right)^4 (\xi_\alpha \xi'_\beta \gamma_{\alpha\beta\mu\nu} \mathcal{E}_{0\mu} \mathcal{E}_{p\nu})^2. \tag{5.197}$$

The factors $\hat{\xi}$ and $\hat{\xi}'$ are the polarization vectors of the incident and final laser light. The electric field from the phonon $\mathcal{E}_p$ is found for wave vector $\vec{q} = \vec{k}_f - \vec{k}_i$, which is sufficiently small that only zone-center phonons are considered. Electric fields are only made by optical modes, and for them the field at small q is

$$\mathcal{E}_{p\mu} = -(4\pi e/V_0)[\hbar/(2\mu\omega_0)]^{1/2}(q_\mu q_\nu/q^2 - \delta_{\mu\nu}/3)\eta_\nu, \tag{5.198}$$

where V_0 is the volume of a unit cell, μ_0 is the reduced mass of the alkali and halide ion pair, ω_0 is the phonon frequency, and η_ν is the polarization component of the phonon.

The theory of FIFORS, when local field effects are included, uses the induced moment given in (5.194). The γ term contributes to the light scattering cross section,

$$\frac{d\sigma}{d\Omega} = \left(\frac{\omega_f}{c}\right)^4 [\gamma_{\alpha\beta\mu\nu} G_{\alpha\alpha'}(\vec{k}_f)\xi_{\alpha'} G_{\beta\beta'}(\vec{k}_i) \times \xi'_{\beta'} \mathcal{E}_{0\nu} G_{\mu\mu'}(\vec{q}) \mathcal{E}_{p\mu'}(\vec{q})]^2, \quad (5.199)$$

which is the same as (5.197) except for the local field corrections. Only the dipole corrections are included, which is accurate for the dielectric function. Then, in the geometry $y(zz)x$ used in the experiments (Krantz and Lüty, 1985), the final expression becomes

$$\frac{d\sigma}{d\Omega} = \left(\frac{\omega_f}{c}\right)^4 [\gamma_1 \mathcal{E}_0 \mathcal{E}_{pz}]^2 \kappa,$$
$$\kappa = \left[\left(\frac{\epsilon_\infty + 2}{3}\right)^2 \left(\frac{\epsilon_0 + 2}{3\epsilon_0}\right)\right]^2 . \quad (5.200)$$

The first factor in the expression for κ comes from the correction to the transverse radiation field, while the second factor comes from the correction to the longitudinal dc field. The value of κ is about 0.9 for the alkali chlorides, and 0.7 for the alkali fluorides. Thus, the local fields make only a small correction to observed cross section in FIFORS.

Values for all the quantities entering (5.200) are known experimentally, except γ_1. The static values, corresponding to $\gamma_1(0; 0, 0, 0)$ have been calculated by TDLDA by Johnson *et al.* (1987), and values relevant for nonlinear refractive index $\gamma_1(-\omega; \omega, \omega, -\omega)$ have been measured by Adair *et al.* (1989) for $\lambda = 1.06\ \mu$m. However, the quantity needed in (5.200) is $\gamma_1(-\omega; \omega, 0, 0)$, for typically, $\lambda = 5145\AA$. In order to evaluate the cross section, we have used the static values calculated by Johnson *et al.* (1987) extrapolated to the FIFORS regime in the manner discussed in Section 5.5 while comparing with the experimental results of Adair *et al.* (1989). In Table 5.18 we give our calculated values for the FIFORS cross section for two potassium halides for which experimental data are available. Also shown are the experimental results of Krantz and Lüty (1985), and the results of Dick (1985) from the phenomenological nonlinear shell model. It is seen that the electro-optic contribution is quite substantial, and the agreement with experiment is reasonable considering the many approximations made, and the absence of any adjustable parameters.

Table 5.18: Electro-optic contribution to field-induced first-order Raman scattering. Available experimental values and the shell model results of Dick (1985) are also shown.

	Cross section (10^{-37}cm^2/sr)		
	EO[a]	Expt.[b]	Shell[c]
KCl	12	32	0.05
KBr	34	46	0.17

[a]Present results. Static, TDLDA results of Johnson *et al.* (1987) for γ, extrapolated to the experimental frequency was used

[b]Krantz and Lüty (1985)

[c]Shell-model result of Dick (1985)

The symmetry of the Raman scattering here is determined by the tensor $\gamma_{\alpha\beta\mu\nu}$. The four indices must come in pairs for the tensor to have non-vanishing values. As already noted, there are only two independent components for cubic symmetry, $\gamma_{zzzz} = \gamma_1$, and $\gamma_{xxzz} = \gamma_{xzxz} = \gamma_2$. In the $y(zz)x$ geometry considered above the wave vector change of the light is in the xy plane; the theory described above predicts there is no electro-optic contribution from the LO mode. Krantz and Lüty (1985) reported the intensity for the LO mode in KCl to be half that for the TO mode; however, in KBr, where they reported a much stronger TO signal than in KCl, they did not detect an LO signal. Thus, Subbaswamy and Mahan (1985) argued that the reported LO signal in KCl might be an overestimate, and that the electro-optic contribution was dominant.

Further, the theory predicts that in a geometry such as $y(zy)x$ with the dc field along z, or $y(zz)x$ with the dc field along x or y, the electro-optic contribution should be approximately equal, and have roughly a tenth of the cross section for the geometry discussed above. This is based on the observation that in this geometry the cross section is determined by γ_2, which is roughly one-third of γ_1. There appear to be no detailed experiments with which to compare these predictions.

Note that the discussion above has been in terms of the local field at the ions, and not in terms of the macroscopic field. In descriptions of Raman scattering in terms of polarizability derivatives with respect to macroscopic fields (see, e.g., Burstein *et al.*, 1968), there is no electro-optic contribution to FIFORS from the TO phonon since there is no macroscopic electric field generated by the TO phonon. By contrast, in our microscopic theory, the local electric field generated by the TO phonon has an appreciable value at the site of the anion, and gives the dominant contribution. The theories based on macroscopic fields pick up this contribution in the lattice displacement derivative of the dielectric constant. Thus, there is no error in the macroscopic theories, but the physics of the coupling with light is somewhat obscured.

5.7.3 Second-Order Raman Scattering

The basis for calculating the electro-optic contribution to second-order Raman scattering in the alkali halides is (5.194). As already mentioned, only the last two terms contribute to Raman scattering. Mahan and Subbaswamy (1986) assumed that only the anion polarizability need be included, which is a reasonable assumption.

In the γ term in (5.194) each phonon makes a field $\mathcal{E}$, and the third field is from the laser. In the B term, the laser makes the $\mathcal{E}$ field, and the two phonons make the quadrupole field Q. If an ion of charge Z_i gets displaced from equilibrium by X_i, there is a dipole $Y_i = Z_i X_i$ and a quadrupole $U_i = Z_i X_i^2$. These dipoles and quadrupoles contribute to the field at ion j according to

$$\begin{aligned} \mathcal{E}^{(j)} &= -\Phi^{(jk)} Z_k (X_k - X_j) + \frac{1}{2}\Psi^{(jk)} Z_k (X_k - X_j)^2 + \cdots, \\ Q^{(j)} &= -\Psi^{(jk)} Z_k (X_k - X_j) + \frac{1}{2}\Xi^{(jk)} Z_k (X_k - X_j)^2 + \cdots, \end{aligned} \quad (5.201)$$

where the quantities ϕ, Ψ, Ξ are the derivatives of the Coulomb potential, defined earlier. The sum of the two phonon wave vectors is such that $\vec{q}_1 + \vec{q}_2 = \vec{k}_f - \vec{k}_i \approx 0$. The result for the cross section includes a summation over the phonon wave vector $\vec{q}$ in the Brillouin zone, and summation over the two branch indices λ and λ'. The expression appropriate to low temperature is

$$\begin{aligned} \frac{d^2\sigma}{d\Omega d\omega} &= \frac{1}{2}\left(\frac{\omega_f}{c}\right)^4 V_0 \int \frac{d^3q}{(2\pi)^3} \sum_{\lambda\lambda'} \delta[\omega_f - \omega_i - \omega_\lambda(\vec{q}) - \omega_{\lambda'}(-\vec{q})] \\ & \left[\xi_\alpha \xi'_\beta \left\{ \gamma_{\alpha\beta\mu\nu} \sum_{\sigma,\sigma'=+,-} G^\sigma_{\mu\mu'}(\vec{q}) \mathcal{E}^\sigma_{p\mu'}(q\lambda) G^{\sigma'}_{\nu\nu'}(-\vec{q}) \mathcal{E}^{\sigma'}_{p\nu'}(-q\lambda') \right.\right. \end{aligned}$$

$$+ \frac{1}{3} B_{\alpha\beta\mu\nu} \Xi_{\mu\nu\mu'\nu'} M_{\mu'\nu'\delta\delta'} U_{\delta\delta'}(0) + p_{\alpha\beta\mu\nu} Q_{\mu}(\vec{q}\lambda) Q_{\nu}(-\vec{q}\lambda') \Big\} \Big]^2 . \tag{5.202}$$

Here, V_0 is the scattering volume, the sum over σ is for the two types of ions, and

$$U_{\delta\delta'}(\vec{q}) = \sum_j \exp{(i\vec{q} \cdot \vec{r}_j)} Z_j X_{j\delta} X_{j\delta'}, \tag{5.203}$$

where the summation is over all ions, Z_j is the ionic charge, and X_j the relative displacement of the jth ion with respect to the anion at the origin.

The last term in (5.202) is the atomic displacement contribution (see, e.g., Hardy and Karo, 1979). As already mentioned, there are no *ab initio* evaluations of this tensor to date. The discussion here will be limited to assessing the importance of the electro-optic contribution alone. At finite

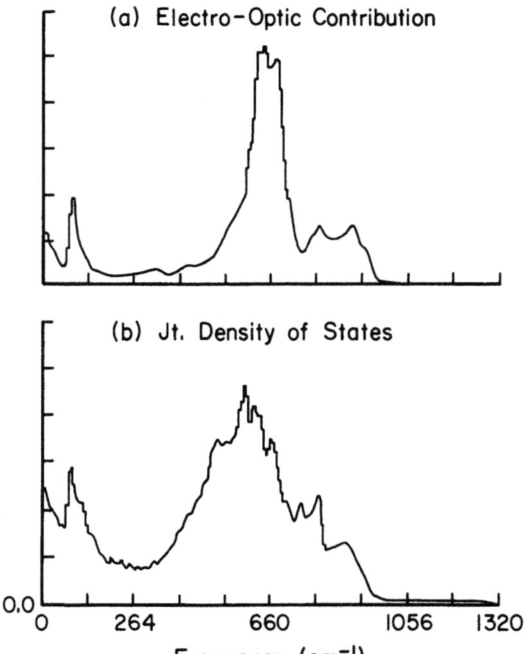

Figure 5.8: (a) Electro-optic contribution to second-order Raman scattering in LiF, as calculated by Mahan and Subbaswamy (1986). (b) The joint density of states, which is very close to the experimental spectral shape.

temperatures one also has to include the standard phonon occupation factors as well as phonon difference processes.

Some details of further simplifications of the above expression for the cross section may be found in the paper by Mahan and Subbaswamy (1986). The method for evaluating the Green's function G may be found in Mahan (1990). Mahan and Subbaswamy evaluated the electro-optic contribution for LiF in the $y(zz)x$ geometry, using the calculated values of γ and B reported by Fowler and Madden (1984). Their calculated spectrum, along with the joint density of states (which very closely resembles the experimental spectrum of Sharma *et al.*, 1973), is shown in Figure 5.8. The electro-optic contribution is large near the TO contributions and small in the region of acoustic phonons, as might be expected. Mahan and Subbaswamy further evaluated the contributions of the γ terms and the B terms separately, and concluded that the B contributions dominate in the case of LiF for the assumed values for the parameters.

Chapter 6

Other Solids and Simple Liquids

6.1 Semiconductors

Given the technological importance of semiconductors, a great deal of first-principles work has been done for these systems ever since the early days of local density computations. The first-principles calculations of dielectric response in semiconductors can be classified into three categories: (i) TDLDA calculations; (ii) frozen-field calculations; and (iii) self-energy corrected TDLDA.

Since our previous discussion has not concerned TDLDA calculations involving explicit band structures, we briefly present the form which takes account of Bloch 's theorem. Recall from Section 2.4 that the TDLDA response function χ is written as

$$\chi^{-1} = \chi_{KS}^{-1} - f_{xc} - |\vec{r} - \vec{r}'|^{-1}, \tag{6.1}$$

where,

$$f_{xc} \approx \frac{d^2}{dn^2}[n\epsilon_{xc}(n)], \tag{6.2}$$

with ϵ_{xc} being the exchange-correlation density for the homogeneous electron gas (see (2.126)). Thus, the dielectric matrix is given by

$$\epsilon^{-1} = 1 + (|\vec{r} - \vec{r}'|^{-1} + f_{xc})\chi. \tag{6.3}$$

From translational symmetry one has

$$\epsilon^{-1}(\vec{r},\vec{r}\,') = \frac{1}{\Omega} \sum_{\vec{q},\vec{G},\vec{G}\,'} e^{i(\vec{q}+\vec{G})\cdot\vec{r}} \epsilon^{-1}_{\vec{G},\vec{G}\,'} e^{-i(\vec{q}+\vec{G}\,')\cdot\vec{r}\,'}. \tag{6.4}$$

The macroscopic dielectric constant is defined as

$$\epsilon_0 = 1/\lim_{q\to 0} \epsilon^{-1}_{00}. \tag{6.5}$$

In this formulation, then, what is needed is the response function for the system of independent Kohn-Sham particles,

$$\begin{aligned} \chi_{KS}(\vec{G},\vec{G}\,',\vec{q}) &= \sum_{n,n\,',\vec{k}} \frac{f_{n,\vec{k}}(1-f_{n',\vec{k}+\vec{q}})}{\lambda_{n,\vec{k}} - \lambda_{n\,',\vec{k}+\vec{q}}} \Big(< n,\vec{k} \mid e^{-i(\vec{q}+\vec{G})\cdot\vec{r}} \mid n\,',\vec{k}+\vec{q} > \\ & \quad \times < n\,',\vec{k}+\vec{q} \mid e^{i(\vec{q}+\vec{G}\,')\cdot\vec{r}\,'} \mid n,\vec{k} > + \text{c.c.} \Big). \end{aligned} \tag{6.6}$$

This is the LDA version of the RPA result of Adler (1962) and Wiser (1963).

The first step is to obtain the LDA eigenvalues and eigenfunctions. For semiconductors this is usually done using norm-conserving pseudopotentials in the plane wave basis (see, e.g., Baroni and Resta, 1986). The explicit calculation of χ_{KS} involves Brillouin zone summations, and is quite tedious. Generally, special point summation is utilized (Chadi and Cohen, 1973). In this explicit calculation of χ_{KS}, sum over the unoccupied conduction bands is involved, and the number of such levels used needs to be consistent with the maximum plane wave energy retained in the expansion (Hybersten and Louie, 1986).

Baroni *et al.* (1987) used a band structure version of the modified Sternheimer method. The primary benefit is that the cumbersome sums over conduction bands are avoided. It is the analog for Bloch electrons of the methods for atoms described in Chapter 3. The linear, induced change in density is written as

$$\begin{aligned} & n^{(1)}(\vec{q}+\vec{G}) \\ = & -\frac{4}{\Omega}\sum_{k}\sum_{v,c} \frac{< v,\vec{k} \mid e^{-i(\vec{q}+\vec{G})\cdot\vec{r}} \mid c,\vec{k}+\vec{q} >< c,\vec{k}+\vec{q} \mid V^{(1)}_{SCF} \mid v,\vec{k} >}{\lambda_c(\vec{k}+\vec{q}) - \lambda_v(\vec{k})}, \end{aligned} \tag{6.7}$$

Table 6.1: Macroscopic dielectric constants of semiconductors calculated by various first-principles methods are compared with experimental values.

	Si	Ge	GaAS
TDLDA	12.7[a]	20.7[b]	12.3[a]
TDLDA/SE[c]	11.2	16.5	
Expt.[d]	11.7	15.8	10.9

[a]Baroni and Resta (1986)
[b]Hybersten and Louie (1986)
[c]Levine and Allan (1989, 1990)
[d]Edwards (1985)

where $V_{SCF}^{(1)}$ is the self-consistent, first-order, screened potential as before. The trick now is to express the induced density as a sum over occupied valence bands only,

$$n^{(1)}(\vec{q}+\vec{G}) = -\frac{1}{\Omega}\sum_{\vec{k}}\sum_{v} < v,\vec{k} \mid e^{-i(\vec{q}+\vec{G})\cdot\vec{r}} \mid \widetilde{v,\vec{k}} >, \tag{6.8}$$

where the first-order perturbed Kohn-Sham Bloch function $\mid \widetilde{v,\vec{k}} >$ obeys the modified Sternheimer equation

$$[H_0 - \lambda_v(\vec{k})] \mid \widetilde{v,\vec{k}} > = V_{SCF}^{(1)} \mid v,\vec{k} > . \tag{6.9}$$

Baroni *et al.* (1987) have pointed out that there is a substantial gain in numerical effort in this procedure over the conventional method of solving first for the Kohn-Sham response function.

The TDLDA results for the static dielectric constant of several semiconductors are shown in Table 6.1, along with experimental values. The TDLDA overestimates the static dielectric constant of Si and Ge by more than 10%. This is consistent with the trend found for atoms and ions discussed earlier, although the disagreement is much higher here.

It was generally believed that this overestimate was due to the LDA itself, and that going beyond LDA in density functional theory would improve it (Baroni and Resta, 1986; Hybersten and Louie, 1987). This overestimate in the dielectric constant is related to the well-known underestimate of the band gap in LDA (Hybersten and Louie, 1986). However, a recent analysis by Godby *et al.* (1988), using the so-called GW approximation (Hedin and Lundquist, 1969), showed that replacing the LDA exchange-correlation potential $V_{xc}(r)$ by a nonlocal self energy $\Sigma(\vec{r}, \vec{r}')$ results in a worse value for the dielectric constant, even though the band gap is reproduced rather well. Mahan and Sernelius (1989) have shown that including vertex corrections in the calculation of the self-energy changes the results significantly. They also conclude the 'GW' approximation is poor.

Levine and Allan (1989) have proposed a scheme which improves the agreement significantly. They incorporated self-energy effects into TDLDA in the form of a so-called "scissors operator." This is merely a k-independent shift of the conduction bands, without changing their eigenfunctions (Godby *et al.*, 1988). The amount of shift is determined by comparison with GW calculations. By this scheme, they found agreement with experimental dielectric constant in both Si and Ge to within a few percent. Their values are also shown in Table 6.1.

Kunc and Tosatti (1984) and Resta and Kunc (1986) have proposed direct schemes to calculate the dielctric matrix which are analogous to the frozen-phonon method mentioned in Chapter 5. The idea in this frozen-field or direct approach is to solve for the total energy of the solid in LDA for several values of an applied static electric field. Thus, the response can be obtained not only in the linear regime, but in the nonlinear regime as well. The difficulty, of course, is that the periodicity is lost, in general. If the $\vec{q}$ of the applied field is a rational fraction of a reciprocal lattice vector, then a supercell can be found such that $\vec{q}$ is a reciprocal lattice vector of the supercell, and periodicity is restored. Of course, computational effort is increased significantly. Another method suggested by Resta and Kunc (1986) is the use of a "saw-tooth" potential in the supercell (which corresponds to a periodically repeated capacitor in the crystal) to simulate the effect of a constant electric field. Note that in the linear regime, the frozen-field method agrees with TDLDA, provided both calculations are carried out with the same pseudopotentials, exchange-correlation potentials, and at the same degree of accuracy.

Only a few calculations of nonlinear coefficients have been done for semiconductors. Recently Moss *et al.* (1990) calculated the third-harmonic coefficient $\chi^{(3)}$ for Si, Ge, and GaAs. They evaluated the coefficients using a band structure approach.

6.2 Metal Surfaces

The study of surface excitations of metals has a long and interesting history. These properties have proved to be particularly challenging both experimentally and theoretically. Experimentally, the problems of surface purity and characterization often lead to irreproducible or contradictory results. Theoretically, one is faced at once with the influence of correlations in screening the discontinuous change in the positive background. Local density calculations found their first successes in this context. Detailed reviews of various issues are given by Lang (1973), Feibleman (1982), and Zangwill (1988). Here we limit ourselves to situations where TDLDA has invoked in calculating the response. A more detailed discussion may be found in Liebsch (1988a).

6.2.1 Static Fields

TDLDA is used to calculate the infinitesimal change in the charge density at the surface caused by an infinitesimal electric field. The calculations have been done for static electric fields, and also for ac fields. First we describe the calculation for static fields. As usual, the first step is to calculate the electron density $n(z)$ at the surface in the absence of applied fields. Most of the calculations have been done for a jellium model of a metal, where the positive ions are replaced by a uniform positive background. Self-consistent LDA calculations of this kind were first done by Lang and Kohn (1970). The positive charge was taken to be a step profile $n_0\theta(-z)$ at the origin, where n_0 is the uniform charge density in the bulk solid. The total charge is

$$n_T(z) = n(z) - n_0\theta(-z), \tag{6.10}$$

$$n(z) = \frac{m}{\pi\hbar^2}\sum_{\epsilon_F<\epsilon_k}(\epsilon_F - \epsilon_k)|\psi_k(z)|^2, \tag{6.11}$$

$$0 = -\frac{\hbar^2}{2m}\psi_k''(z) + (V(z) - \epsilon_k)\psi(z). \tag{6.12}$$

The LDA potential $V(z)$ contains the Hartree, Coulomb, and exchange-correlation potentials. The latter contributions are evaluated using LDA.

These calculations seem straightforward. They are similar in spirit to our atomic calculations, since they involve a numerical integration over just one variable. In this case the variable is z. A major numerical problem is the Friedel oscillations at large values of z inside of the solid. The charge density oscillates with a wave vector of $2k_F$, and with a slowly decreasing amplitude. These oscillations make the numerical work difficult. Another feature is that the surface must be constrained to be charge-neutral at each iteration. Simple iterative procedures do not converge well, and various relaxation methods are required to obtain a self-consistent solution.

An infinitesimal electric field $\mathcal{E}$ can be applied to the surface of this jellium solid. The charge density can be expanded in powers of the field

$$n(z, \mathcal{E}) = n(z) + n^{(1)}(z)\mathcal{E} + n^{(2)}(z)\mathcal{E}^2 + \dots . \tag{6.13}$$

The equilibrium electron density, without the field, is $n(z)$. It is the quantity calculated above, and as usual will be written without a superscript. The first-order response $n^{(1)}(z)$ is the first order part of the quantity $\delta n(z, \omega = 0)$ evaluated at zero frequency. It determines the position of the image plane for charges outside of the metal surface. Figure 6.1 shows the equilibrium and first order density profiles at a jellium surface calculated for a metal with $r_s = 4$. These calculations were done by Lang and Kohn (1970) using a self-consistent potential. They are probably the first calculations done using TDLDA. The Friedel oscillations are evident on the left side of the figure. The electron density extends appreciably beyond the jellium edge for a bulk metal surface. The centroid of the induced electron density locates the position of the image plane, within which the average field is essentially zero. It is found to lie a distance z_1 outside the jellium boundary,

$$z_1 = \int_{-\infty}^{\infty} dz\, z\, n^{(1)}(z). \tag{6.14}$$

Lang and Kohn (1970) found from their self-consistent calculation that z_1 is positive, depends on the value of r_s, and is of the order of 1-2 a.u.

Calculations have also been done for the second-order surface charge $n^{(2)}(z)$ in response to an electric field. Numerical convergence problems prevent a direct application of iterative methods. A clever method of obtaining a solution was used by Gies and Gerhardts (1985). Classical electrostatics shows that the total surface charge σ is proportional to the applied electric field $\mathcal{E}$. Gies and Gerhardts solved for the electron density at the surface $n(z, \sigma)$ as a function of the total charge at the surface. This method is numerically much easier than applying an external field. Their results can be expanded

$$n(z, \sigma) = n(z) + n^{(1)}(z)\sigma + n^{(2)}(z)\sigma^2 + \dots . \tag{6.15}$$

They calculate the surface charge profile with a finite density σ of surface charge.

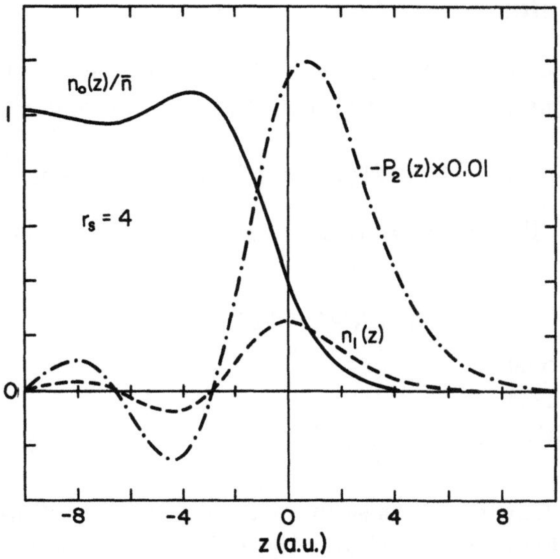

Figure 6.1: Electronic charge density at a jellium surface for a bulk density with $r_s = 4$. Solid line is the equilibrium density $n(z)$, while the dashed line is the linear surface charge $n^{(1)}(z)$. These results were caculated by Lang and Kohn (1970). The third line labeled 'P_2' is the integral of the second-order charge density calculated by Weber and Liebsch (1987).

The first and second order densities $n^{(1,2)}(z)$ in this expansion are directly proportional to the same terms in the field expansion in (6.13). Weber and Liebsch (1987) have used this relation to calculate the second-harmonic response to static fields. Their results are shown in Figure 6.1 as the line labeled 'P_2,' which means the second-order polarization. Using the definition of polarization as $\partial P/\partial z = n(z)$, the second-order polarization is the integral of the second-order density,

$$P_2(z) = \int_{-\infty}^{z} dz' n^{(2)}(z'). \tag{6.16}$$

The value of P_2 vanishes at $z = \pm\infty$ because the area under $n^{(2)}$ is zero. The curve for P_2 has a shape similar to $n^{(1)}(z)$. Both have the feature that the induced charge is largely outside of the metal surface. Similar curves are found for other values of r_s.

6.2.2 ac Fields

For metal surfaces, one is interested in the response of the metal electrons, which occupy the half space $z \leq 0$, to the external electric potential

$$\phi_{ext}(\vec{r}, t) = \phi(\vec{q}, \omega) \exp\left(i\vec{q}_{\|} \cdot \vec{r}_{\|} + qz + i\omega t\right), \tag{6.17}$$

with $q = \mid \vec{q}_{\|} \mid$. The quantity of interest, as before, is the induced electron density:

$$\delta n(z, q, \omega) = n(z, \phi_{ext}) - n_0. \tag{6.18}$$

The linear response is

$$n^{(1)}(z, q, \omega) = \int dz' \chi_1(z, z', q, \omega) \phi_{ext}(z', q, \omega), \tag{6.19}$$

where χ_1 is the density response function of the semi-infinite metal. Usually it is evaluated for q=0. As shown by Feibelman (1982), many measurable quantities can be expressed in terms of the total induced surface charge

$$\sigma(\omega) = \int dz \delta n(z, 0, \omega) = \frac{\epsilon(\omega) - 1}{\epsilon(\omega) + 1}, \tag{6.20}$$

and the induced dipole moment density

$$d(\omega) = \frac{1}{\sigma(\omega)} \int dz\, z\, \delta n(z, 0, \omega). \tag{6.21}$$

For instance, the photoabsorption cross section for p-polarized light is given by (Feibelman, 1982)

$$\frac{d\sigma}{d\omega} = \sqrt{8}\, \frac{e^2}{c\hbar} \omega (1 - \omega^2/\omega_p^2) \Im\{d(\omega)\}, \tag{6.22}$$

where ω_p is the bulk plasmon frequency. The surface plasmon dispersion for long wavelengths is given by (Feibelman, 1982)

$$\omega_s(q) = \frac{\omega_p}{\sqrt{2}} \left(1 - \frac{q}{2} d(\omega_s)\right). \tag{6.23}$$

In addition, the van der Waals interaction between a neutral atom and a metal surface (Zaremba and Kohn, 1976; Liebsch, 1987), the electronic contribution to the lifetime of adsorbate vibrations, and a host of other quantities depend ultimately on $\delta n(z, q, \omega)$. We will discuss the van der Waals interaction in the next subsection.

Formally, the determination of δn for the semi-infinite metal in TDLDA is identical to our earlier discussion in Chapters 2 and 3. Specifically, in the Green's function approach, for long wavelength one has (Liebsch, 1988a)

$$\delta n(z,\omega) = \int dz' \chi_{KS}(z,z',\omega) V_{SCF}(z',\omega), \tag{6.24}$$

where χ_{KS} is the independent particle susceptibility of the ground state obtained in the Kohn-Sham formulation, and V_{SCF} is the screened, self-consistent, perturbing potential consisting of the applied potential, the induced change in the Hartree potential, and the induced change in the exchange-correlation potential.

The technical details of carrying out the TDLDA procedure for the surface problem are substantially different from those we have discussed in earlier chapters. First of all, the geometry and boundary conditions for the ground state problem already are very different, and the Kohn-Sham equations are solved by basis set expansions (see, e.g., Lang, 1973). In carrying out the response calculation in TDLDA, the primary numerical complication is the handling of the Friedel oscillations in the induced density in the interior of the metal, as already mentioned. Several sum rules (Liebsch, 1985; 1988b) can be used to avoid the integration of the induced density in the interior region. Further, separation of the induced Hartree potential into long-range bulk and short-range surface contributions facilitates the imposition of appropriate asymptotic behavior (Liebsch, 1988a). The qualitative differences between LDA and TDLDA response for surface phenomena are similar to those found in the case of free atoms and ionic solids earlier.

As illustration of a frequency-dependent quantity, we show in Figure 6.2 the quantity $\omega(1-\omega^2/\omega_p^2)\Im\{d(\omega)\}$, which is proportional to photoabsorption cross section for p-polarized light, calculated by Liebsch (1988a) for $r_s = 4$. Results for other values of r_s are similar. The solid line is the TDLDA result, and the dashed line is the LDA result. One notes the resonance peak is shifted to lower frequency by inclusion of exchange and correlation in the screening cloud. Weber and Liebsch (1987) and Liebsch and Schaich (1989) performed calculations relevant to nonlinear response at metal surfaces. In particular, they calculated the second-harmonic charge density $n^{(2)}(z,2\omega)$, which is defined as the density which oscillates at a frequency of 2ω. It is given by a combination of the second-order susceptibility χ_2 plus a term from the second-order potential

$$n^{(2)}(z,2\omega) = \int dz' \int dz'' \chi_2(z,z',z'',\omega,\omega) V_{SCF}^{(1)}(z',\omega) V_{SCF}^{(1)}(z'',\omega)$$

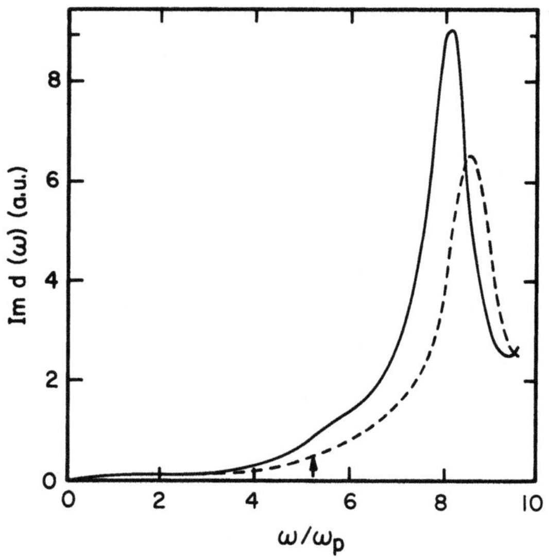

Figure 6.2: Photoabsorption cross section from a metal surface for p-polarized light ; results r_s = 4 are shown. The solid line is TDLDA, while the dashed line is LDA result. From Liebsch (1988a).

$$
\begin{aligned}
& + \int dz' \chi_1(z, z', 2\omega) V^{(2)}_{SCF}(z', \omega), \\
V^{(1)}_{SCF}(z, \omega) &= \phi_{ext}(z, q, \omega) - 4\pi \int dz' |z - z'| n^{(1)}(z', \omega) \\
& + n^{(1)}(z, \omega) \frac{\partial}{\partial n} V_{xc}(n), \\
V^{(2)}_{SCF}(z, \omega) &= -4\pi \int dz' |z - z'| n^{(2)}(z', 2\omega) + n^{(2)}(z, 2\omega) V'_{xc} \\
& + \frac{1}{2} [n^{(1)}(z, \omega)]^2 V''_{xc}.
\end{aligned}
$$

Another parameter which can be calculated is $a(\omega)$, which controls the efficiency of second-harmonic generation. It is given by

$$
a(\omega) = \frac{4n_0}{\sigma(\omega)^2} \int z dz n^{(2)}(z, \omega). \tag{6.25}
$$

Calculations of these quantities are given by Liebsch and Schaich (1989). Complex behavior is found as a function of z and ω. The experimental situation has not yet advanced to the stage where reliable values for comparison with theory are available.

6.2.3 Surface van der Waals Interaction

In Section 4.3 the van der Waals interaction between atoms was discussed. When a neutral atom is outside of the surface of a solid, the long-range interaction between the atom and the surface is due to van der Waals interaction. When the solid surface is locally planar, the nonrelativistic form for the interaction is:

$$V(z) = -\frac{C_3}{(z-z_0)^3}, \tag{6.26}$$

$$C_{3,i} = \frac{1}{4\pi}\int_0^\infty du\alpha_{d,i}(iu)\frac{\epsilon(iu)-1}{\epsilon(iu)+1}, \tag{6.27}$$

$$z_0 = \frac{1}{4\pi C_3}\int_0^\infty du\,\alpha_{d,i}(iu)\frac{\epsilon(iu)(\epsilon(iu)-1)}{(\epsilon(iu)+1)^2}\int_{-\infty}^\infty zdz\delta n(z,iu).$$

The formula for C_3 is due to Lifshitz (1956). The formula for the displacement z_0 of the image plane is due to Zaremba and Kohn (1976). The term z_0 has another contribution due to surface currents, which is usually neglected. The solid is characterized by the dielectric function $\epsilon(iu)$ evaluated at imaginary frequency, and by the induced surface charge $\delta n(z,iu)$. The latter quantity has been calculated for real frequency by Schaich and Kempa (1987).

As an example, consider simple metals where the dielectric response function at long wavelength is dominated by the plasma frequency ω_p. Then one finds that $\epsilon(iu) = 1 + \omega_p^2/u^2$. The integral for C_3 is evaluated using this expression for $\epsilon(iu)$ and the London formula (4.71) for $\alpha_{d,i}(iu)$, to get

$$C_{3,i} = \frac{1}{8}\alpha_{d,i}(0)\frac{E_i E_s}{E_i + E_s}. \tag{6.28}$$

The energy of the surface plasmon is $E_s = \hbar\omega_p/\sqrt{2}$. This simple formula is useful for making estimates of the van der Waals constant for simple metals. Unfortunately, most experimental results are available for the noble metals or the transition metals. The dielectric function $\epsilon(\omega)$ for these metals is more complicated than for the simple metals. Table 6.2 shows some values of C_3 calculated by Bruch (1983). The first column denotes the molecules, while the first row denotes the different metal surfaces. The last surface is

Table 6.2: The coefficient C_3 in units of $e^2a_0^2$. The left-hand column denotes the gas atom or molecule, while the top row denotes the metal surface. 'Gr' denotes the basal plane of graphite. The values for H_2 are averaged over the orientations of the molecule. From Bruch (1983).

	Cu	Ag	Au	Al	Pd	Gr
H	0.124	0.126	0.135	0.123	0.112	0.090
H_2	0.169	0.174	0.188	0.166	0.152	0.121
He	0.058	0.062	0.068	0.054	0.052	0.043
Ne	0.121	0.130	0.143	0.110	0.108	0.086
Ar	0.402	0.420	0.458	0.384	0.362	0.30
Kr	0.569	0.602	0.642	0.546	0.512	0.43
Xe	0.841	0.871	0.941	0.812	0.758	0.61

graphite. The only metal on the list which is simple is aluminum. Using its surface plasmon energy of $E_s = 0.80$ Rydbergs in formula (6.28) gives values for C_3 about 10% smaller than those listed by Bruch. The simple formula works well in this case.

Attempts to measure the van der Waals potential on neutral atoms or molecules outside of metal surfaces have not been completely successful. Lando and Slutsky (1970) found that the force law did not go as z^{-3}. Shih and Parsegian (1975) found the z^{-3} law was obeyed, but the coefficient C_3 disagreed significantly with the theoretical value. This subject seems to be an untidy corner of physics where experiment disagrees with the theory. A thorough discussion is given by Mehl and Schaich (1977).

6.3 Small Metal Particles

In recent years there has been a great deal of experimental work on the electronic properties of microscopic metallic particles. One motivation comes from the desire to understand how a crystal grows from the aggregation of atoms. Another comes from the possibility of producing new materials with small clusters as the building blocks. A recent review of metal cluster physics has been given by de Heer *et al.* (1987).

One of the striking features of small metal clusters is that the cluster size dependence of stability, polarizability, etc. exhibits features associated with distinct shell structure such as is observed in the case of isolated atoms and nuclei (see, e.g., Knight *et al.*, 1984). Such clusters have been successfully modeled in terms of jellium spheres. The ground state properties as well as the response of these systems have been studied extensively by the methods described in this book by various groups, and we refer the reader to the review by de Heer *et al.* (1987) for references.

The effect of the ion cores is taken as a uniform positive charge n^+ equal to that of the electron density in the bulk; the radius of the jellium sphere with Z total valence electrons is

$$R = Z^{1/3} r_s, \tag{6.29}$$

where r_s is the usual Wigner-Seitz electron radius. The sphere is treated in the local density functional formalism described in earlier chapters, with the nuclear Coulomb potential replaced by that due to the positive background charge. The spherical symmetry simplifies the problem considerably, just as in the atomic case. Strong binding occurs for "magic numbers" of electrons, 8, 18, 20, 34, 40, etc., corresponding to closed shells. The classical result for the polarizability of the jellium sphere is simply

$$\alpha_{cl} = R^3, \tag{6.30}$$

where R is the jellium radius. The semi-classical or quantum mechanical results for the static polarizability may be written as

$$\alpha = (R + \delta)^3, \tag{6.31}$$

where δ will be called the polarizability parameter. It is evident that for large R, δ should approach the result for the shift of the image plane z_1 for a semi-infinite jellium discussed in the previous section.

There have been several TDLDA calculations of the polarizability of jellium spheres, based on modified Sternheimer procedure (Beck, 1984), as well as the Green 's function procedure (Ekardt, 1984; Puska *et al.* 1985), all of which are in agreement. Beck 's treatment is very similar to that of Mahan (1980c) for rare gas atoms. The numerical procedure used by Beck to solve the inhomogeneous Schrödinger equation is somewhat different. Recall that our procedure, described in Chapter 3, is a Numerov procedure. Beck uses a matrix formulation to carry this out.

A detailed comparison of the size dependence of the static polarizability as a function of cluster size was made by Ekardt (1985a). While nonself-consistent calculations predict that the sphere polarizability decreases from the classical value as the radius decreases, TDLDA predicts, first of all, a slower dependence on the size, and secondly, an increase from the classical value as the radius decreases. Ekardt (1985a) noted that the r dependence of the induced density across the sphere radius was almost identical for all sphere sizes, explaining the slow dependence of α on cluster size. Secondly, the dependence of the induced density across the sphere boundary was very similar to its dependence across the boundary for a semi-infinite jellium as calculated by Lang and Kohn (1973). In the latter case, the observed static polarizability is greater than the classical value (positive z_1, as described above). Thus, in the absence of self-consistent screening (or in the presence of an artificial infinite barrier used in some calculations), the induced density has a centroid at a radius below that of the sphere, giving a smaller polarizability. Experiments on alkali metal clusters by Knight *et al.* (1985) confirm the slow decrease of polarizability with increasing cluster size. Thus, TDLDA gives the correct physics at relatively little computational effort for jellium spheres.

Ekardt (1985a) also calculated the photoabsorption cross section, which is proportional to the imaginary part of the dynamic polarizability (see (4.22)). Most of the calculational details are the same as in the atomic case. A typical result is shown in Figure 6.3. The TDLDA result clearly shows all the classical and quantum effects — cusps due to particle-hole excitations, the Mie resonance, and the Landau-damped bulk plasmon.

In a subsequent paper Ekardt (1985b) considered the collective, multipole excitations in jellium spheres. These calculations are similar to that of the quadrupole and octupole polarizabilities of atoms and ions considered in previous chapters. Mahan (1980) found that the effect of screening is less important for higher multipole response. The reason appears to be the rather highly inhomogeneous density response for the higher angular momentum components. In an analogous manner, Ekardt (1985b) found that the screening vanishes rather abruptly at a critical value of angular momentum l^{cr} of the multipole field. Ekardt showed that this value is approximately equal to $q_{\parallel}^{cr}/R$, where $q_{\parallel}^{cr}$ is the critical value of wave vector at which surface plasmon at a planar jellium surface is very heavily damped, and R is the radius of the sphere. We are aware of no experimental measurement of l-pole excitations in metal clusters.

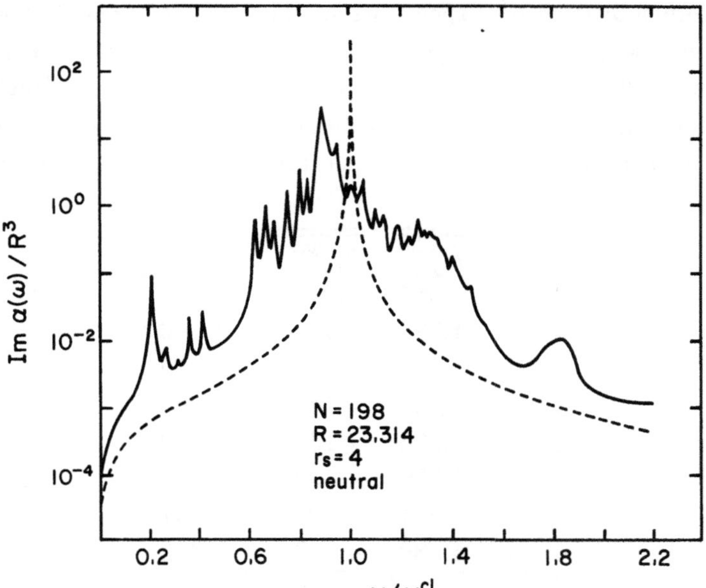

Figure 6.3: The calculated photoabsorption cross section for a jellium sphere for a 198 atom cluster at r_s=4. The solid line is TDLDA, and the dashed line is LDA result. From Ekardt (1985a).

6.4 Rare Gas Solids

In Chapter 4 polarizabilities and hyperpolarizabilities of free rare gas atoms were presented. In Chapter 5 the use of the spherical solid model for alkali halide crystals was described. Using similar methods Senatore and Subbaswamy (1986) calculated the density dependence of the polarizability of solid argon. Their results compare favorably with pressure experiments of Sinnock and Smith (1969) as described below. They used the self-consistent spherical solid model described in Section 5.1. No pseudopotentials are available for rare gas atoms. Thus, the potential due to the neighbors is constructed from a superposition of electron density of the neighbors, suitably spherically averaged (see (5.22)-(5.26)). Senatore and Subbaswamy (1986) used the Ceperley-Alder correlation. The calculation of the polarizability follows exactly the same procedure as in the case of alkali halide crystals. The polarizability was calculated for several different values of the lattice constant, corresponding to different pressures. In order to compare with the

Table 6.3: Density dependence of the Lorentz-Lorenz factor in solid argon. The first column is density ρ in units of 10^{-2} mole/cm^3. The next two columns are the Lorentz-Lorenz function in units of cm^3/mole. The last column is the ratio of the calculated to the experimental quantity. This ratio shows a constant error of 4%.

ρ	$F_{LL}^{cal,a}$	$F_{LL}^{exp,b}$	$F_{LL}^{cal}/F_{LL}^{exp}$
~ 0	4.365	4.217	1.035
4.093	4.321	4.156	1.040
4.170	4.311	4.148	1.039
4.235	4.307	4.144	1.039
4.293	4.303	4.138	1.040
4.343	4.299	4.132	1.040
4.385	4.294	4.124	1.041
4.416	4.289	4.117	1.042

[a] G. Senatore and K.R. Subbaswamy, Phys. Rev. B **34**, 5754 (1986)

[b] A.C. Sinnock and B.L. Smith, Phys. Rev. **181**, 1297 (1969)

measured parameters, they calculated the "Lorentz-Lorenz function "

$$F_{LL} = \frac{1}{\rho}\frac{n^2-1}{n^2+2} = \frac{4\pi N_A \alpha}{3}. \tag{6.32}$$

where N_A is Avogadro's number. Their calculated values are shown in Table 6.3. The first row lists the values for free atoms. The remaining rows show the values at the indicated density. The last column shows the ratio of the calculated to the experimental quantity. There is a constant error of 4%. We have mentioned previously that the TDLDA method has this error. This calculation shows that TDLDA can be used to calculate the density dependence of the polarizability. As the density increases, the atom is compressed and the polarizability decreases.

6.5 Simple Liquids

The microscopic treatments of simple liquids, such as rare gases and molten binary salts, have generally proceeded from effective interatomic potentials and numerical simulations. They treat the atoms and ions as point particles for the most part, and occasionally as point-polarizable particles. While this is often adequate for the calculation of thermodynamic properties and neutron structure factors, it proves inadequate for the description of light scattering and certain other optical properties.

The theoretical treatment of the dielectric properties of a liquid is much more involved, in general, than that in a crystal. The recent scheme proposed by Car and Parrinello (1985), which essentially combines LDA and molecular dynamics, offers a way out at the price of a computational effort that in principle should not be too much greater than for a crystal.

These computational schemes certainly constitute state-of-the-art tools when dealing with condensed media. Yet, there can be situations in which different and much simpler schemes may be fruitfully employed. In simple liquids such as rare gas liquids, where the electrons are well localized in the vicinity of the nuclear centers, the method used for the crystalline case in the preceding chapter can be used to take account of the electronic response in the liquid in a more realistic manner. We have seen that the self-consistent spherical solid model gives good results for the polarizability of rare gas atoms in the crystalline environment. Johnson *et al.* (1987) showed that the same method could be readily applied to treat disordered systems, provided that the pair correlation function $g(r)$ is given.

In considering the crystalline case, the requirement that all atoms in the assembly should be equivalent was used to impose the self-consistency condition $n_j(\vec{r}) = n_0(\vec{r} - \vec{R}_j)$, with $\vec{R}_j$ the position of the jth atom relative to the *central* one (see (5.24) and its discussion). The same condition can be imposed in the liquid as an average property.

Before presenting the results obtained by Johnson *et al.* (1987) for the Lorentz-Lorenz function of liquid rare gases with this approach, we examine the the accuracy of this scheme for describing the ground state. Johnson *et al.* calculated the cohesive energy of rare gas crystals in this scheme, and thereby, deduced the Lennard-Jones potential parameters appropriate to the liquid density.

The ground state calculation for the crystal has already been described in Section 5.1. Since the local density approximation does not account for the correlation of nonoverlapping electrons, total energy calculations for the crystal were performed retaining only the first shell of neighbors. This should yield that part of the cohesive energy that originates in the overlap region and that should be reasonably described by LDA. In Table 6.4 the cohesive energies per atom obtained in this way, at the experimental lattice spacing,

Table 6.4: Cohesive energy contributions (Rydberg/atom) for rare gas crystals. The LDA values were calculated with nearest-neighbor overlaps. The last column gives the contribution from second farther neighbors calculated using Lennard-Jones pair potentials with empirical parameters from McDonald and Singer (1972).

	E'^{a}_{LDA}	$E^{stat\,b}_{exp}$	E''_{LJ}
Ne	0.0015	0.0019	0.0005
Ar	0.0039	0.0065	0.0019
Kr	0.0051	0.0090	0.0028
Xe	0.0078	0.0126	0.0039

[a]Johnson *et al.* (1988)
[b]Pollack (1964)

are compared with the *static lattice* experimental energies.

The contribution to the cohesive energy from second and farther neighbors, which are not included in LDA, was estimated using Lennard-Jones pair potentials with empirical parameters given by McDonald and Singer (1972). These are also shown in Table 6.4. Johnson *et al.* found that the discrepancies are reasonably well explained in terms of the interaction with further neighbors, which is out of the reach of LDA.

Johnson *et al.* also performed a series of calculations at different values of the lattice parameter, around the minimum of the calculated cohesive energy of the crystalline cluster. These results were fit to a Lennard-Jones (LJ) pair potential form according to

$$E'_{LDA}(R_1) = 2N_1\epsilon\left[\left(\frac{\sigma}{R_1}\right)^{12} - \left(\frac{\sigma}{R_1}\right)^{6}\right]. \tag{6.33}$$

Here, N_1 and R_1 are, respectively, the number of nearest neighbors and their distance from the *central* atom. The estimates of σ and ϵ that one obtains are sensitive to the range of R_1's used (Johnson *et al.*, 1988). To reduce the arbitrariness of the fitting procedure, Senatore *et al.* (1988) chose a range of distances of 1 Bohr radius, and determined its position with respect to the minimum of E'_{LDA} by minimizing the discrepancies between the fit and the actual LDA values.

The LJ effective potentials that one extracts in this manner contain in an average fashion contributions that go beyond the two-body interactions.

Table 6.5: Lennard-Jones parameters extracted from LDA are compared with empirical values.

		Ne	Ar	Kr	Xe
σ(Å)	LDA[a]	2.86	3.42	3.66	3.92
	Emp.[b]	2.76	3.40	3.63	3.95
ϵ (10^{-14} erg)	LDA[a]	0.53	1.41	1.86	2.83
	Emp.[b]	0.50	1.62	2.25	3.12

[a]Senatore *et al.* (1988)
[b]McDonald and Singer (1972)

As such, they should not be compared with genuine two-body potentials between rare gas atoms, as extracted from molecular beam scattering experiments for instance. Rather they should be compared with empirical LJ potentials. This comparison is made in Table 6.5. On the whole, the quality of the LDA Lennard-Jones parameters appears to be fair. It is at least comparable to that obtained by *APW*-Xα calculations (Trickey *et al.*, 1974), if not better. Thus, the LDA scheme is at least capable of giving a semiquantitative description of rare gases in the condensed phase.

As mentioned above, to apply the LDA scheme to the liquid phase the pair correlation function $g(r)$ is needed. In the absence of accurate experimental results for rare gas liquids, Johnson *et al.* used the molecular dynamics results of Verlet (1968) for a Lennard-Jones fluid. In doing so they used the empirical LJ parameters obtained by McDonald and Singer (1972) by a global fit of liquid properties on a large range of thermodynamic parameters. This should provide a reliable description of the two-body correlations in the rare gas liquids.

Using these parameters and the triple-point thermodynamic data of Crawford (1974) one finds that for Ar, Kr, and Xe the reduced temperature $T^* \approx 0.72$ and the reduced density $\rho^* \approx 0.85$. Hence, Johnson *et al.* used Verlet's tabulated values for $g(r)$ at $T^* = 0.72$ and $\rho^* = 0.85$, which are the closest to the actual values.

In performing the ground state calculation for the liquid, only the effect of the first shell of neighbors on the *central* atom was kept, since only the

Table 6.6: Lorentz-Lorenz function F_{LL} (cm^3 mol^{-1}) for rare gas liquids at the triple point, and fractional change with respect to free atom values.

	Calculated[a]		Expt.[b]	
	F_{LL}^{liq}	Δ	F_{LL}^{liq}	Δ
Ar	4.428	-1.1	4.217	-1.1
Kr	6.574	-2.4	6.453	0.5
Xe	10.503	-2.1	10.475	-0.1

[a]Johnson *et al.* (1988)
[b]Sinnock and Smith (1969)

nearest neighbor overlap should be important. Thus, the integral over the interatomic distance in performing the spherical averaging (see (5.25)) was cut off at the first minimum of $r^2g(r)$ (i.e., r=1.44 in reduced units). This corresponds to a coordination number in the liquid of 11.72, to be compared with 12 in the fcc crystal.

In Table 6.6 the calculated and the experimental Lorentz-Lorenz functions (see (6.32)) for the liquid at triple point are presented. Also shown are the fractional changes from the atomic value,

$$\Delta = F_{LL}^{liq}/F_{LL}^{at} - 1. \tag{6.34}$$

The theoretical results are for the static case, whereas the experimental data were measured at a wavelength of 5461 Å for the liquids, and 5463 Å for the rarified gases. However, since dispersion is very weak this should not cause any serious discrepancy. In comparing the calculated atomic polarizability with the measured Lorentz-Lorenz function, it is assumed that, once the local field effects are taken into account at the point dipole level, the only remaining effects arise from the the overlap compression of the atom by its neighbors.

The agreement between theory and experiments is reasonable as far as the absolute value of the polarizability is concerned, in that the quality of the agreement in the liquid phase is similar to that for isolated atoms . (In fact, the overlap compression in the liquid has only a very small effect due to the weakness of interactions in these closed shell systems.) As noted in Chapter 4, the agreement could be somewhat improved by considering a modification of LDA in which self-interaction corrections are included. The theory predicts in all cases a decrease of polarizability, in accord with simple considerations on overlap compression. It also gives a larger effect in the heavier atoms. This can be understood by invoking the greater softness of larger atoms, which makes them more sensitive to the environment. Experiments fully confirm these predictions in the case of Ar. They suggest, on the contrary, much smaller effects for Kr and Xe, and even give the opposite sign for Kr. It is not easy to understand an increase in polarizability in going from the free atom situation to the condensed phase. Sinnock and Smith (1968) have suggested that this inconsistency might well be due to inaccuracies in the gas data.

Bibliography

Chapter 1

Ahlberg, R., and O. Goscinski, *J. Phys. B* **6**, 2254 (1973)

Buckingham, A.D., *Adv. Chem. Phys.* **12**, 107 (1967)

Fock,V., *Z. Phys.* **61**, 126 (1930)

Hartree, D.R., *Proc. Cambridge Phil. Soc.* **24**, 89 (1928)

Hohenberg, P., and W. Kohn, *Phys. Rev.* **136**, B864-871 (1964)

Kohn, W., and L.J Sham, *Phys. Rev.* **140**, A1133-1138 (1965)

Lundqvist, S., and N.H. March, *Theory of the Inhomogeneous Electron Gas* (Plenum, 1983)

Mahan, G.D., *Many-Particle Physics* 2nd edition (Plenum, 1990)

Slater, J.C., *Phys. Rev.* **81**, 385 (1951); **82**, 538 (1951); **91**, 52 (1953)

Slater, J.C., *Solid-State and Molecular Theory* (Wiley, 1975)

Wigner, E.P., *Trans. Faraday Soc.* **34**, 678 (1938)

Chapter 2

Ahlberg, R., and O. Goscinski, *J. Phys. B* **6**, 2254 (1973)

Ashcroft, N.W., and D. Mermin, *Solid State Physics* (Holt, Rinehart, and Winston, 1976)

Bartolatti, L.J., *Phys. Rev. A* **24**, 1661 (1981); **26**, 2243 (1982)

Bloembergen, N., *Nonlinear Optics* (Benjamin, 1965)

Buckingham, A.D., *Adv. Chem. Phys.* **12**, 107 (1967)

Callaway, J., and N.H. March, Solid State Phys. **38**, 135 (1984)

Chakravarty, S., M.B. Fogel, and W. Kohn, *Phys. Rev. Lett.* **43**, 775 (1979)

Devreese, J.T., V.E. VanDoren, and P.E. VanCamp, eds., *Ab initio Calculations of Phonon Spectra* (Plenum, 1983)

DeCicco, P.D., and F.A. Johnson, *Proc. Roy. Soc. London A* **310**, 111 (1969)

Dirac, P.A.M., *Proc. Cambridge Phil. Soc.* **26**, 376 (1930)

Feynman, R. P., *Phys. Rev.* **36**, 340 (1939)

Flytzanis, Ch., in *Quantum Electronics: A Treatise*, Vol. 1, edited by H. Rabin, C.L. Tang, and N. Bloembergen (Academic, 1975), Ch. 2

Fowler, P.W., and P.A. Madden, *Phys. Rev. B* **29**, 1035 (1984)

Fowler, P.W., and P.A. Madden, *Phys. Rev. B* **31**, 5443 (1985)

Frankel, J., *Wave Mechanics* (Oxford, 1934)

Gross, E.K.U., and W. Kohn, *Phys. Rev. Lett.* **55**, 2850 (1985)

Hohenberg, P., and W. Kohn, *Phys. Rev.* **136**, B864-871 (1964)

Jones, R.O., and O. Gunnarsson, *Rev. Mod. Phys.* **61**, 689 (1989)

Kohl, H., and R.M. Driezler, *Phys. Rev. Lett.* **56**, 1993 (1986)

Kohn, W., and L.J Sham, *Phys. Rev.* **140**, A1133-1138 (1965)

Kugler, A.A., *J. Stat. Phys.* **12**, 35 (1975)

Langhoff, P.W., S.T. Epstein, and M. Karplus, *Rev. Mod. Phys.* **44**, 602 (1972)

Lundqvist, S., and N.H. March, *Theory of the Inhomogeneous Electron Gas* (Plenum, 1983)

March, N.H., *Self-Consistent Fields in Atoms* (Oxford, 1975)

Mahan, G.D., *Phys. Rev. A* **22**, 1780 (1980)

Mahan, G.D., *Many-Particle Physics*, 2nd edition (Plenum, 1990)

Ng, T.K., and K.S. Singwi, *Phys. Rev. Lett.* **59**, 2627 (1987)

Runge, E., and E.K.U. Gross, *Phys. Rev. Lett.* **52**, 997 (1984)

Seitz, F., *Modern Theory of Solids* (McGraw-Hill, 1940)

Sewell, G.L., *Proc. Cambridge Phil. Soc.* **45**, 678 (1949)

Shen, Y.R., *The Principles of Nonlinear Optics* (Wiley, 1984)

Singhal, S.P., and J. Callaway, *Phys. Rev. B* **14**, 2347 (1976)

Stott, M.J., and E. Zaremba, *Phys. Rev. A* **21**, 12 (1980)

von Weizsäcker, C.F., *Z. Phys.* **96**, 431 (1935)

Waller, I., *Z. Phys.* **38**, 635 (1926)

Xu, B.X., and A.K. Rajagopal, *Phys. Rev. A* **31**, 2682 (1985)

Zangwill, A., and P. Soven, *Phys. Rev. A* **21**, 1561 (1980)

Chapter 3

Arfken, G., *Mathematical Methods for Physicists* (Academic, 1970)

Atkinson, K.E., *An Introduction to Numerical Analysis* (Wiley, 1978)

Callaway, J., and N.H. March, Solid State Phys. **38**, 135 (1984)

Carr, W.J., R.A. Coldwell-Horsfall, and A.E. Fein, *Phys. Rev.* **124**, 747 (1961)

Carr, W.J., and A.A. Maradudin, *Phys. Rev.* **133**, A371 (1964)

Ceperley, D.M., and B.J. Alder, *Phys. Rev. Lett.* **45**, 566 (1980)

Eaves, J.O., and S.T. Epstein, *J. Chem. Phys.* **61**, 3486 (1974)

Gordon, R.G., and Y.S. Kim, *J. Chem. Phys.* **56**, 3122 (1972)

Gunnarsson, O., and B.I. Lundqvist, *Phys. Rev.* B **13**, 4274 (1976)

Herman, F., and S. Skillman, *Atomic Structure Calculations* (Prentice-Hall, 1963)

Hubbard, J., *Proc. Roy. Soc. London A* **243**, 336 (1957)

Lindgren, I., *Int. J. Quantum Chem.* **5**, 411 (1971)

Mahan, G.D., *Phys. Rev. A* **22**, 1780 (1980)

Perdew , J.P., and A. Zunger, *Phys. Rev. B* **23**, 5048 (1981)

Sakurai, J.J., *Modern Quantum Mechanics* (Addison-Wesley, 1985)

Senatore, G., and K.R. Subbaswamy, *Phys. Rev. A* **35**, 2440 (1987)

Singwi, K.S., and A. Sjölander (1967, unpublished)

Singwi, K.S., and M.P. Tosi, Solid State Phys. **36**, 177 (1981)

Sternheimer, R.M., *Phys. Rev.* **96**, 951 (1954)

Stott, M.J., and E. Zaremba, *Phys. Rev. A* **21**, 12 (1980); (E) **22**, 2293 (1980)

Vashishta, P., and K.S. Singwi, *Phys. Rev. B* **6**, 875 (1972)

Zangwill, A., *J. Chem. Phys.* **78**, 5926 (1983)

Zangwill, A., and P. Soven, *Phys. Rev. A* **21**, 1561 (1980)

Chapter 4

Ahlberg, R., and O. Goscinski, *J. Phys. B* **8**, 2149 (1975)

Almbladh, C.-O., and U. von Barth, *Phys. Rev. B* **31**, 3231 (1985)

Amusia, M. Ya., *Phys. Lett.* **14**, 36 (1963)

Ashcroft, N.W., and D. Mermin, *Solid State Physics* (Holt, Rinehart, and Winston, 1976)

Axilrod, B.M., and E. Teller, *J. Chem. Phys.* **11**, 299 (1943)

Bartolotti, L.J., *J. Chem. Phys.* **80**, 5687 (1984)

Bell, R.J., and I.J. Zucker, in *Rare Gas Solids*, ed. M.L. Klein and J.A. Venables (Academic, 1976) Ch. 2

Bloembergen, N., *Nonlinear Optics* (Benjamin, 1965)

Böbel, G., A. Longinetti, and F. Fumi, *J. Phys. (Paris)* **48**, 45 (1987)

Brandt, W., and B. Lundqvist, *Phys. Rev.* **132**, 2135 (1963)

Bruch, L.W., *Surf. Sci.* **125**, 194 (1983)

Buckingham, A.D., *Adv. Chem. Phys.* **12**, 107 (1967)

Dalgarno, A., *Adv. Chem. Phys.* **12**, 143 (1967)

Doren, M.B., *J. Phys. B* **7**, 558 (1974)

Fowler, P.W., and P.A. Madden, *Phys. Rev. B* **30**, 6131 (1984)

Fowler, P.W., and P.A. Madden, *Phys. Rev. B* **31**, 5443 (1985)

Glover, R.M., and F. Weinhold, *J. Chem. Phys.* **65**, 4913 (1976)

Langhoff, P.W., and R.P. Hurst, *Phys. Rev.* **139**, A1415 (1965)

Leonard, P.J., and J.A. Barker, *Theoretical Chemistry, Advances and Perspectives*, Vol. 1, ed. H. Eyring and D. Henderson (Academic, 1975) p. 117

Levine, Z., and P. Soven, *Phys. Rev. A* **29**, 625 (1984)

Lombardi, J.C., *Phys. Rev. A* **36**, 1445 (1987)

London, F., *Z. Phys. Chem. B* **11**, 222 (1930); *Z. Phys.* **63**, 245 (1930); *Trans. Faraday Soc.* **33**, 8 (1936)

Mahan, G.D., *J. Chem. Phys.* **43**, 1569 (1965)

Mahan, G.D., *Phys. Rev. A* **22**, 1780 (1980)

Mahan, G.D., *J. Chem. Phys.* **76**, 493 (1982)

Mahan, G.D., *Many-Particle Physics*, 2nd edition (Plenum, 1990)

Marr, G.V., and J.B. West, *At. Data Nucl. Data Tables* **18**, 497 (1976)

Rehr, J.J., E. Zaremba, and W. Kohn, *Phys. Rev. B* **12**, 2062 (1975)

Rotenberg, M., R. Bivins, N. Metropolis, and T.K. Wooten, Jr., *The 3-j and 6-j Symbols* (Technology Press, 1959)

Schiff, L.I., *Quantum Mechanics*, Third Edition (McGraw-Hill, 1968)

Schmidt, P.C., A. Weiss, and T.P. Das, *Phys. Rev. B* **19**, 5525 (1979)

Senatore, G., and K.R. Subbaswamy, *Phys. Rev. A* **34**, 3619 (1986)

Senatore, G., and K.R. Subbaswamy, *Phys. Rev. A* **35**, 2440 (1987)

Senatore, G., and K.R. Subbaswamy (unpublished, 1990)

Shelton, D.P., *J. Chem. Phys.* **84**, 404 (1986)

Stott, M., and E. Zaremba, *Phys. Rev. A* **21**, 12 (1980); (E) **22**, 2293 (1980)

Stott, M., E. Zaremba, and D. Zobin, *Can. J. Phys.* **60**, 210 (1982)

Subbaswamy, K.R., and G.D. Mahan, *J. Chem. Phys.* **84**, 15 (1985)

Tang, K.T., J.M. Norbeck, and P.R. Certain, *J. Chem. Phys.* **64**, 3063 (1976)

Teachout, R.R., and R.T. Pack, *At. Data* **3**, 195 (1971)

Wendin, G., *Structure and Bonding* **45** (Springer-Verlag, 1981), p.1

Wendin, G., *Comments Mod. Phys. D* **17**, 115 (1986)

Zangwill, A., *J. Chem. Phys.* **78**, 5926 (1983)

Zangwill, A., and P. Soven, *Phys. Rev. A* **21**, 1561 (1980)

Zaremba, E., and W. Kohn, *Phys. Rev. B* **13**, 2270 (1976)

Chapter 5

Adair, R., L.L. Chase, and S.A. Payne, *Phys. Rev. B* **39**, 3337 (1989)

Aggarwal, K.G., and B. Szigeti, *J. Phys. C* **3**, 1097 (1970)

Akhmanov, S.A., L.B. Meisner, S.T. Parinov, S. M. Saltiel, and V.G. Tunkin, *Zh. Eksp. Teor. Fiz.* **73**, 1710 (1977) [*Sov. Phys. - JETP* **46**, 898 (1977)]

Almbladh, C.O., and U. von Barth, *Phys. Rev. B* **13**, 3307 (1976)

Baroni, S., P. Gianozzi, and A. Testa, *Phys. Rev. Lett.* **58**, 1861 (1987)

Baroni, S., and R. Resta, *Phys. Rev. B* **33**, 5969 (1986)

Bilz, H., and W. Kress, *Phonon Dispersion Relations in Insulators* (Springer-Verlag, 1979)

Born, M., and M. Bradburn, *Proc. Roy. Soc. London A* **188**, 161 (1946)

Bryant, G.W., *Phys. Rev. B* **19**, 2801, 2864 (1979)

Bryant, G.W., and G.D. Mahan, *Phys. Rev. B* **17**, 1744 (1978)

Buckingham, A.D., *Adv. Chem. Phys.* **12**, 107 (1967)

Burstein, E., in *Dynamical Processes in Solid State Optics*, edited by R. Kubo and H. Kamimura (Benjamin, 1966), p. 34

Burstein, E., and P.L. Smith, *Phys. Rev.* **74**, 229 (1948)

Car, R., and M. Parrinello, *Phys. Rev. Lett.* **22**, 2471 (1985)

Cochran, W., *CRC Rev. Solid State Sci.* **2**, 1 (1971)

Cohen, R.E., L.L. Boyer, and M.J. Mehl, *Phys. Rev. B* **35**, 5749 (1987)

Colella's group (1982; 1985): D.R. Yoder and R. Colella, *Phys. Rev. B* **25**, 2545 (1982); M.S. Schmidt, R. Colella, and D.R. Yoder-Short, *Acta Cryst. A* **41**, 171 (1985)

Dick, B.G., and A.W. Overhauser, *Phys. Rev.* **112**, 90 (1958)

Dick, B.G., *Phys. Stat. Solidi (B)* **127**, 131 (1985)

Elliott, D.S., and J.F. Ward, *Mol. Phys.* **51**, 45 (1984)

Fischer, K., H. Bilz, R. Haberkorn, and W. Weber, *Phys. Stat. Solidi (B)* **54**, 285 (1972)

S. Gironcoli, K., S. Baroni, and R. Resta, *Phys. Rev. Lett.* **62**, 2853 (1989)

Gordon, R.G., and Y.S. Kim, *J. Chem. Phys.* **56**, 3122 (1972)

Gunnarsson, O., J. Harris, and R.O. Jones, *J. Chem. Phys.* **67**, 3970 (1977)

Hans, D., A.A.S. Sangachin, and M. Kumar, *Phys. Stat. Solidi (B)* **147**, 529 (1988)

Hardy, J.R., *Phil. Mag.* **4**, 1278 (1959); **7**, 315 (1962)

Hardy, J.R., and A.M. Karo, *The Lattice Dynamics and Statics of Alkali Halide Crystals* (Plenum, 1979)

Hayes, W., and R. Loudon, *Scattering of Light by Crystals* (Wiley, 1978)

Heine, V., and I.V. Abarenkov, *Phil Mag.* **9**, 451 (1964); **12**, 529 (1965)

Hellwarth, R.W., *Prog. Quantum Electron.* **5**, 1 (1977)

Hybersten, M.S., and S.G. Louie, *Phys. Rev. B* **35**, 5585 (1987)

Jennison, D.R., and A.B. Kunz, *Phys. Rev. B* **13**, 5597 (1976)

Johnson, M.D., K.R. Subbaswamy, and G. Senatore, *Phys. Rev. B* **36**, 9202 (1987)

Johnson, M.D., K.R. Subbaswamy, and G. Senatore, *Phys. Rev. B* **37**, 6508 (1988)

Johnson, M.D., and K.R. Subbaswamy, *Phys. Rev. B* **39**, 10280 (1989)

Karo, A.M., and J.R. Hardy, *Phys. Rev.* **129**, 2024 (1963)

Kirkpatrick, S., C.D. Gelatt, Jr., and M.P. Vecchi, *Science* **220**, 671 (1983)

Krantz, M., and F. Lüty, *Phys. Rev. B* **31**, 2599 (1985)

Krishnan, R. S., T.N. Vasudevan, and P. S. Narayanan, *Proceedings of the Nuclear Physics and Solid State Physics Symposium, Bombay, December 27-31, 1974*, **17**C, 198

Kunc, K., M. Balkanski, and M.A. Nusimavici, *Phys. Rev. B* **12**, 4346 (1975)

Landolt-Börnstein Numerical Data and Functional Relationships in Science and Technology – New Series, Part III, Vol. **11** (1978)

Levine, B.F., *Phys. Rev. B* **7**, 2600 (1973)

Loudon, R., *Proc. Roy. Soc. London A* **188**, 161 (1946)

Lowndes, R.P., and D.H. Martin, *Proc. Roy. Soc. London A* **308**, 473 (1969)

Mahan, G.D., *Phys. Rev.* **153**, 983 (1967)

Mahan, G.D., *Solid State Ionics* **1**, 29 (1980)

Mahan, G.D., *Solid State Commun.* **33**, 797 (1980a)

Mahan, G.D., *Phys. Rev. A* **22**, 1780 (1980b)

Mahan, G.D., *Phys. Rev. B* **34**, 4235 (1986); (E) *B* **38**, 7841 (1988)

Mahan, G.D., and M. Mostoller, *Phys. Rev. Lett.* **57**, 357 (1986); *Phys. Rev. B* **34**, 5726 (1986); *Phys. Rev. B* **41** 10840 (1990)

Mahan, G.D., and K.R. Subbaswamy, *Phys. Rev. B* **33**, 8657 (1986)

Maradudin, A.A., E.W. Montroll, G.H. Weiss, and I.P. Ipatova, *Theory of Lattice Dynamics in the Harmonic Approximation* (Academic, 1971)

Mayer, J.E., *J. Chem. Phys.* **1**, 270 (1933)

Mizrahi, V., and D.P. Shelton, *Phys. Rev. Lett.* **55**, 696 (1985)

Moyer, C.A., *Phys. Rev. B* **9**, 3555 (1974)

Nye, J.F., *Physical Properties of Crystals* (Oxford, 1957)

Pauling, L., *Proc. Roy. Soc. London A* **114**, 181 (1927)

Pauling, L., *The Nature of the Chemical Bond*, 3rd Ed. (Cornell, Ithaca, 1960)

Petrashen', M.I., I.V. Aberenkov, and N.N. Kristofel', *Opt. Spectrosc.* **9**, 276 (1961)

Pettersen, H.E., *J. Phys. Chem. Solids* **42**, 1027 (1981)

Pettersson, S., *J. Phys.: Condensed Matter* **1**, 347 (1989)

Pettersson, S., and K.R. Subbaswamy, *Phys. Rev. B* (1990, in press)

Phillips, J.C., *Bands and Bonds in Semiconductors* (Academic Press, 1973)

Ruffa, A.R., *Phys. Rev.* **130**, 1412 (1963)

Sangster, M.J.L., *J. Phys. Chem. Solids* **34**, 355 (1967)

Schröder, U., *Solid State Commun.* **4**, 347 (1966)

Senatore, G., and K.R. Subbaswamy, *Phys. Rev. B* **34**, 5754 (1986)

Senatore, G., M.D. Johnson, and K.R. Subbaswamy, *Phil. Mag. A* **58**, 53 (1988)

Sharma, T.P., R.D. Kirby, and S.S. Jaswal, *Phys. Rev. B* **9**, 1971 (1973)

Sharma, H.P., J. Shanker, and M.P. Verma, *J. Phys. Chem. Solids* **37**, 1077 (1976)

Subbaswamy, K.R., and G.D. Mahan, *Phys. Rev. B* **32**, 5453 (1985)

Szigeti, B., *Proc. Roy. Soc. London A* **204**, 51 (1950)

Tessman, J.R., A.H. Kahn, and W. Shockley, *Phys. Rev.* **92**, 890 (1953)

Van Vechten, J.A., *Phys. Rev.* **182**, 891 (1969)

Vedam, K., and E.D.D. Schmidt, *Solid State Commun.* **3**, 373 (1965)

Vedam, K., and E.D.D. Schmidt, *J. Mat. Sci.* **1**, 310 (1966)

Vedam, K., , E.D.D. Schmidt, J.L. Kirk, and W.C. Schneider, *Mat. Res. Bull.* **4**, 573 (1969)

Von der Lage , F.C., and H. Bethe, *Phys. Rev.* **71**, 612 (1947)

Wendel, H., and R.M. Martin, *Phys. Rev. B* **19**, 5251 (1979)

Wigner, E., and F. Seitz, *Phys. Rev.* **43**, 804 (1933)

Wilson, J.N., and R.M. Curtis, *J. Phys. Chem.* **74**, 187 (1970)

Woodruff, T.O., *Solid State Commun.* **46**, 139 (1983)

Yin, M.T., and M.L. Cohen, *Phys. Rev. Lett.* **45**, 1004 (1980)

Zaremba, E., L.M. Sander, H.B. Shore, and J.H. Rose, *J. Phys. F* **9**, 1763 (1977)

Chapter 6

Adler, S.L., *Phys. Rev.* **126**, 413 (1962)

Baroni, S., and R. Resta, *Phys. Rev. B* **33**, 7017 (1986)

Baroni, S., P. Gianozzi, and A. Testa, *Phys. Rev. Lett.* **58**, 1861 (1987)

Beck, D., *Phys. Rev. B* **30**, 6935 (1984)

Bruch, L.W., *Surf. Sci.* **125**, 194 (1983)

Car, R., and M. Parrinello, *Phys. Rev. Lett.* **22**, 2471 (1985)

Chadi, D.J., and M.L. Cohen, *Phys. Rev. B* **8**, 5747 (1973)

Chou, M.Y., and M.L. Cohen, *Phys. Rev. Lett.* **52**, 2141 (1984)

Crawford, R.K., in *Rare Gas Solids*, Vol. 2, edited by M.L. Klein and J.A. Venables (London, Academic Press, 1976), Ch. 11

de Heer, W.A., W.D. Knight, M.Y. Chou, and M.L. Cohen, Solid State Phys. **40**, 93 (1987)

D.F. Edwards, in *Handbook of Optical Constants of Solids*, E.D. Palik, ed. (Academic, 1985)

Ekardt, W., *Phys. Rev. B* **29**, 1558 (1984)

Ekardt, W., *Phys. Rev. B* **31**, 6360 (1985)

Ekardt, W. , *Phys. Rev. B* **32**, 1961 (1985)

Feibelman, P.J., *Phys. Rev. B* **12**, 1319 (1982)

Feibelman, P.J., *Prog. Surf. Sci.* **12**, 287 (1982)

Gies, P., and R.R. Gebhardts, *Phys. Rev. B* **31**, 6843 (1985)

Godby, R.W., M. Schlüter, and L.J. Sham, *Phys. Rev. B* **37**, 10159 (1988)

Hedin, L., and S. Lundqvist, Solid State Phys. **23**, 1 (1969)

Hybersten, M.S., and S.G. Louie, *Phys. Rev. B* **34**, 5390 (1986)

Hybersten, M.S., and S.G. Louie, *Phys. Rev. B* **35**, 5585 (1987)

Johnson, M.D., K.R. Subbaswamy, and G. Senatore, *Phys. Rev. B* **37**, 6508 (1988)

Knight, W.D., K. Clemenger, W.A. de Heer, W.A. Saunders, *Phys. Rev. B* **31**, 2539 (1985)

Kunc, K., and E. Tosatti, *Phys. Rev. B* **29**, 7045 (1984)

Lando, D., and L.J. Slutsky, *J. Chem. Phys.* **52**, 1510 (1970)

Lang, N. D., Solid State Phys., **28**, 225 (1973)

Lang, N.D., and W. Kohn, *Phys. Rev. B* **1**, 4555 (1970)

Levine, Z., and D.C. Allan, *Phys. Rev. Lett.* **63**, 1719 (1989)

Levine, Z., and D.C. Allan, *Phys. Rev. B* (1990) (to be published)

Liebsch, A., *Phys. Rev. Lett.* **54**, 67 (1985)

Liebsch, A., *Phys. Rev. B* **35**, 9030 (1987)

Liebsch, A., in *Electrodynamics of Interfaces and Composite Media*, edited by R.G. Berrera and W.L. Mochán (World Scientific, 1988a), p.16

Liebsch, A., *Phys. Rev. Lett.* **61**, 1233 (1988b)

Liebsch, A., and W.L. Schaich, *Phys. Rev. B* **40**, 5401 (1989)

Lifshitz, I.M., *Sov. Phys. JETP* **2**, 73 (1956)

Mahan, G.D., *Phys. Rev. A* **22**, 1780 (1980)

Mahan, G.D., and B. Sernelius, *Phys. Rev. Lett.* **62**, 2718 (1989)

McDonald, I.R., and K. Singer, *Mol. Phys.* **23**, 29 (1972)

Mehl, M.J., and W.L. Schaich, *Phys. Rev. A* **16**, 921 (1977); *Surf. Sci.* **99**, 553 (1980)

Moss, D.J., E. Ghahramarni, J.E. Sipe, and H.M. Van Driel, *Phys. Rev. B* **41**, 1542 (1990)

Pollack, G.L., *Rev. Mod. Phys.* **36**, 748 (1964)

Puska, M.J., R.M. Nieminen, and M. Manninen, *Phys. Rev. B* **31**, 3487 (1985)

Rehr, J.J., E. Zaremba, and W. Kohn, *Phys. Rev. B* **12**, 2062 (1975)

Resta, R., and K. Kunc, *Phys. Rev. B* **34**, 7146 (1986)

Sangster, M.J.L., *J. Phys. Chem. Solids* **34**, 355 (1967)

Schaich, W.L., and K. Kempa, *Phys. Scripta* **35**, 204 (1987)

Senatore, G., and K.R. Subbaswamy, *Phys. Rev. B* **34**, 5754 (1986)

Senatore, G., M.D. Johnson, and K.R. Subbaswamy, *Phil. Mag.* **58**, 53 (1988)

Shih, A., and V.A. Parsegian, *Phys. Rev. A* **12**, 835 (1975)

Sinnock, A.C., and B.L. Smith, *Phys. Rev.* **181**, 1297 (1969)

Trickey, S.B., F.R. Green, and F.W. Averill, *Phys. Rev. B* **8**, 4822 (1973)

Verlet, L., *Phys. Rev.* **165**, 201 (1968)

Weber, M., and A. Liebsch, *Phys. Rev. B* **35**, 7411 (1987)

Wiser, N., *Phys. Rev.* **129**, 62 (1963)

Zangwill, A., *The Physics of Surfaces* (Oxford University Press, 1988)

Zaremba, E., and W. Kohn, *Phys. Rev. B* **13**, 2270 (1976)

Index

GPSR Compliance

The European Union's (EU) General Product Safety Regulation (GPSR) is a set of rules that requires consumer products to be safe and our obligations to ensure this.

If you have any concerns about our products, you can contact us on ProductSafety@springernature.com

In case Publisher is established outside the EU, the EU authorized representative is:

Springer Nature Customer Service Center GmbH
Europaplatz 3
69115 Heidelberg, Germany

Batch number: 10615115

Printed by Printforce, the Netherlands